Springer Series in Computational Physics

Springer Series in Computational Physics

A Computational Method in Plasma Physics
F. Bauer, O. Betancourt, P. Garabedian

Implementation of Finite Element Methods for Navier-Stokes Equations
F. Thomasset

Finite-Difference Techniques for Vectorized Fluid Dynamics Calculations
Edited by D. Book

Unsteady Viscous Flows. D. P. Telionis

Computational Methods for Fluid Flow. R. Peyret, T. D. Taylor

Computational Methods in Bifurcation Theory and Dissipative Structures
M. Kubicek, M. Marek

Optimal Shape Design for Elliptic Systems. O. Pironneau

The Method of Differential Approximation. Yu. I. Shokin

Computational Galerkin Methods. C. A. J. Fletcher

Numerical Methods for Nonlinear Variational Problems
R. Glowinski

Numerical Methods in Fluid Dynamics. Second Edition
M. Holt

Computer Studies of Phase Transitions and Critical Phenomena
O. G. Mouritsen

Finite Element Methods in Linear Ideal Magnetohydrodynamics
R. Gruber, J. Rappaz

Numerical Simulation of Plasmas. Y. N. Dnestrovskii, D. P. Kostomarov

Computational Methods for Kinetic Models of Magnetically Confined Plasmas
J. Killeen, G. D. Kerbel, M. C. McCoy, A. A. Mirin

Spectral Methods in Fluid Dynamics. Second Edition
C. Canuto, M. Y. Hussaini, A. Quarteroni, T. A. Zang

Computational Techniques for Fluid Dynamics 1. Second Edition
Fundamental and General Techniques. C. A. J. Fletcher

Computational Techniques for Fluid Dynamics 2. Second Edition
Specific Techniques for Different Flow Categories. C. A. J. Fletcher

Methods for the Localization of Singularities in Numerical Solutions of Gas Dynamics Problems. E. V. Vorozhtsov, N. N. Yanenko

Classical Orthogonal Polynomials of a Discrete Variable
A. F. Nikiforov, S. K. Suslov, V. B. Uvarov

Flux Coordinates and Magnetic Field Structure: A Guide to a Fundamental Tool of Plasma Theory
W. D. D'haeseleer, W. N. G. Hitchon, J. D. Callen, J. L. Shohet

Monte Carlo Methods in Boundary Value Problems. K. K. Sabelfeld

Optimization of Weighted Monte Carlo Methods. G. A. Mikhailov

Gennadii A. Mikhailov

Optimization of Weighted Monte Carlo Methods

Translated by
Karl K. Sabelfeld

Springer-Verlag
Berlin Heidelberg New York London Paris
Tokyo Hong Kong Barcelona Budapest

Gennadii A. Mikhailov

Computing Center, Sibirian Division, USSR Academy of Science,
Prospekt Akademika M. A. Lavrentyeva 6, 630090 Novosibirsk, USSR

Translator

Karl K. Sabelfeld

Computing Center, Sibirian Division, USSR Academy of Science,
Prospekt Akademika M. A. Lavrentyeva 6, 630090 Novosibirsk, USSR

Title of the original Russian edition: *Optimizatsiya Vesovykh metodov Monte-Karlo*
© Nauka, Moscow 1987

ISBN 3-540-53005-3 Springer-Verlag Berlin Heidelberg New York
ISBN 0-387-53005-3 Springer-Verlag New York Berlin Heidelberg

Typesetting: camera ready by author
57/3140 – 5 4 3 2 1 0 – Printed on acid-free paper

Preface

The Monte Carlo method is based on the numerical realization of natural or artificial models of the phenomena under considerations. In contrast to classical computing methods the Monte Carlo efficiency depends weakly on the dimension and geometric details of the problem. The method is used for solving complex problems of the radiation transfer theory, turbulent diffusion, chemical kinetics, theory of rarefied gases, diffraction of waves on random surfaces, etc. The Monte Carlo method is especially effective when using multi-processor computing systems which allow many independent statistical experiments to be simulated simultaneously.

The weighted Monte Carlo estimates are constructed in order to diminish errors and to obtain dependent estimates for the calculated functionals for different values of parameters of the problem, i.e., to improve the functional dependence. In addition, the weighted estimates make it possible to evaluate special functionals, for example, the derivatives with respect to the parameters.

There are many works concerned with the development of the weighted estimates. In Chap. 1 we give the necessary information about these works and present a set of illustrations. The rest of the book is devoted to the solution of a series of mathematical problems related to the optimization of the weighted Monte Carlo estimates.

In Chap. 2 the importance simulation technique is described. In particular, we explain how to use the asymptotics of the radiative transfer problems for improving the corresponding weighted estimates. The main and most interesting chapters are Chaps. 3–6, which are based on the author's own work. In these chapters the nonlinear and minimax theories of weighted Monte Carlo estimates, the vector weighted estimates and the randomized algorithms are presented. Chapter 7 deals with the splitting method. It should be noted that up until now there has been a paucity of works concerned with this important and universal weighted variance reduction method.

In Chap. 8 the relations between transformations of the integral equations and the effectiveness of the weighted methods are derived.

In the Appendix we present the most common simulation methods for sampling random variables and vectors. The repetition method is described here in detail for constructing random vectors with non-Gaussian distributions.

Novosibirsk, August 1991 *Gennadii A. Mikhailov*

Preface

Acknowledgements

The author is grateful to many colleagues for their kind help. He especially thanks Professor K.K. Sabelfeld for his participation in the writing of Sects. 1.1 and 1.4, which deal with the Monte Carlo methods for solving boundary integral equations of the potential theory.

Contents

1. Mathematical Models of Weighted Monte Carlo Methods

1.1 Simple Facts from Functional Analysis

This book deals mainly with optimization of weighted Monte Carlo methods for solving integral equations of the second kind:

$$\boxed{\varphi(x) = \int_X k(x,x')\varphi(x')\,dx' + h(x)} \tag{1.1}$$

or in vector form

$$\varphi = K\varphi + h \quad ,$$

where X is an n-dimensional Euclidian space which will sometimes be called a phase space. The equation (1.1) will be solved, as a rule, in the space L_p, i.e., in the space with the norm

$$\|h\| = \operatorname*{vrai\,sup}_x \|h(x)\| < \infty, \quad K \in [L_\infty \to L_\infty] \quad , \tag{1.2}$$

and

$$\|K\| = \operatorname*{vrai\,sup}_x \int_X k(x,x')|\,dx', \tag{1.3}$$

under the condition that the kernel $k(x,x')$ is a measurable function. The relation (1.3) holds, e.g., if $k(x,x') \geq 0$. In the general case

$$\|K\| \leq \operatorname*{vrai\,sup}_x \int_X |k(x,x')|\,dx'.$$

For simplicity we shall omit the symbol vrai and the expression "almost everywhere with respect to Lebesgue measure" if a misunderstanding of the utilization of the space L_∞ will not arise. The kernel $k(x,y)$ can be quite arbitrary, but it is necessary to satisfy the condition that $K \in [L_\infty \to L_\infty]$; in particular, the kernel can include delta-functions as in the case of transfer theory [1.1]. Then the integration in (1.1) is carried out in fact over a set whose dimension is less than that of X.

If $k(x,y) \geq 0$, then it is sometimes convenient to consider (1.1) in the cone of nonnegative functions. In Chap. 5 an equation for the covariance matrix of the weighted vector estimate will be considered in a more complicated cone of functions whose values lie in the space of positive definite matrices. The corresponding operators will be positive. These operators will be used only in Chap.5; therefore we shall give the information needed there. Here we just note that a cone is defined as a convex set which includes the zero element and posesses the following property: if x belongs to the cone then $-x$ does not belong to the cone.

If the kernel is regular (the corresponding conditions are well-known, [1.2], the integral operator K is compact.

The spectral radius of the operator K is often used in investigating the accuracy and the computational cost of the weighted Monte Carlo methods and is defined as follows [1.2]:

$$\rho(K) = \lim_{n\to\infty} \|K^n\|^{1/n}, \quad \text{and} \quad \rho(K) \leq \|K^n\|^{1/n}, \quad n = 1, 2, \ldots \quad .$$

If the operator K is compact and in the invariant subspaces corresponding to eigenvalues whose moduli are equal to $\rho(K)$ there are no additional elements, then the following inequality holds [1.2]:

$$\|K^n\| \leq c[\rho(K)]^n, \quad n = 1, 2, \ldots \quad . \tag{1.4}$$

In the general case [1.2]

$$\|K^n\| \leq c(\epsilon)[\rho(K) + \epsilon]^n, \quad \forall \quad \epsilon > 0.$$

Since the condition which ensures that these inequalities hold is quite general, we will use them in Chaps. 2 and 5 without special references.

If $\rho(K) < 1$, then the solution to (1.1) can be represented in the form of a convergent Neumann series:

$$\varphi = \sum_{n=0}^{\infty} K^n h. \tag{1.5}$$

Simple estimations of the solution $\varphi(x)$ to (1.1) with nonnegative entries can be obtained on the basis of (1.5).

Lemma 1.1. Let $k(x,x') \geq 0, \quad h(x) \geq 0, \quad \psi(x) \geq 0, \quad \rho(K) < 1$. Then from

$$\psi \leq K\psi + h, \quad \psi \geq K\psi + h$$

the inequalities

$$\varphi \geq \psi, \quad \varphi \leq \psi$$

follow.

Proof. For an arbitrary function $\varphi_0 \in L_\infty$ we have

$$\varphi = \lim_{n\to\infty} \varphi_n, \quad \varphi_n = K\varphi_{n-1} + h = \sum_{m=0}^{n-1} K^m h + K^n \varphi_0 .$$

From this, taking $\varphi_0 = \psi$, we get the proof since the integral operator with a nonnegative kernel is monotone. □

In Sect. 2.5 the following property of a special nonlinear operator related to the operator K will be used.

Lemma 1.2. Let $q = \|K\| < 1$, $h(x) \geq 0$. Then there exists a unique in L_∞ solution g^* to the equation

$$g = G(g) \quad ,$$

where $G(g) = [(Kg)^2 + h]^{1/2}$, and $g^* \geq 0$.

Proof. Let $g_1, g_2 \in L_\infty$. Since h is nonnegative, we get

$$\begin{aligned}|G(g_1) - G(g_2)| &\leq | \, |Kg_1| - |Kg_2| \, | \\ &\leq |Kg_1 - Kg_2| = |K(g_1 - g_2)| \leq q \|g_1 - g_2\|,\end{aligned}$$

which ensures the existence and uniqueness of the solution $g^* \in L_\infty$. Since $G(g) \geq 0$ for all $g \in L_\infty$, we get $g^* \geq 0$. □

In Sect. 1.3 we show that the integral equations of transfer theory have a form adjoint to (1.1) while the importance functions satisfy (1.1). Note that the linear functionals of the transfer theory can be written in the form

$$(f, \varphi) = \int_X f(x)\varphi(x)\, dx \quad , \tag{1.6}$$

where $f \in L_1$ is the source distribution density. Therefore it is sufficient, in the transfer theory, to optimize esimates for the function $\varphi(x)$ (Note that this optimization is uniform with respect to x, see Chaps. 2–5). Our experience shows that utilization of (1.6) essentially simplifies the investigation and optimization of the weighted Monte Carlo algorithms for calculating the linear functionals of the transfer theory.

The relation (1.5) can be rewriten as follows:

$$\varphi = [I - K]^{-1} h,$$

where $[I - K]^{-1}$ is the resolvent of K (or the resolvent operator).

In this book we consider mainly the algorithms based on the utilization of the Neumann series, i.e., the resolvent is considered only in the neighbourhood of the origin $\lambda = 0$. This approach is not effective if the Neumann series converges slowly and even fails if it is neceessary to solve the equation

$$\varphi(x) = \lambda K\varphi + h \tag{1.7}$$

at $\lambda = \lambda^* : |\lambda^*| \geq |\lambda_1|$, where λ_1 is the characteristic number of (1.7) having minimal modulus. In this case the Neumann series diverges. Thus two problems can then be formulated:

(1) accelerate the convergence of the slow convergent series $\varphi = R_{\lambda^*}h$ in the case when λ^* is close to λ_1 $[\lambda^* \lesssim \lambda_1]$, where $R_\lambda = I + \lambda K + \lambda^2 K^2 + \ldots$ is the resolvent.

(2) If $|\lambda^*| \geq |\lambda_1|$, construct the resolvent R_λ (which is in the case of compact operator K a meromorphic function of λ) using only the iterations $K^n h$.

This formulation was first given by *Sabelfeld* in [1.3], where Monte Carlo algorithms for solving boundary integral equations of the potential theory were constructed. In these cases it is necessary to solve the second problem . As mentioned in [1.3], the choice of the method for calculating the resolvent depends on the information about the eigenvalues of the integral equation. Here we consider only the method of analytical continuation of the Neumann series using transformation of the spectral parameter $\lambda = \psi(\eta)$. This method is described in detail in [1.3].

Let λ_k be the poles of the meromorphic function R_λ which are the characteristic numbers of (1.7). Consider the construction of the solution to this equation at $\lambda \in D$, $\lambda \neq \lambda_k$ $(k = 1, 2, \ldots)$, where D is a simple-connected domain lying inside of the domain of definition of the function R_λ such that D includes the point $\lambda = 0$. It is supposed that in the neighbourood of the origin the function is given by

$$R_\lambda h = \sum_{n=0}^{\infty} c_n \lambda^n, \quad c_n = K^n h. \tag{1.8}$$

In the complex plane of the variable η we choose a simple-connected domain Δ (as a rule, a unit disk) including the point $\eta = 0$, and construct a function $\psi(\eta)$ which carries out a conformal mapping of Δ to the domain D. Then the representation

$$R_\eta = \sum_{n=0}^{\infty} b_n \eta^n \tag{1.9}$$

holds, which is obtained by substituting $\lambda = \psi(\eta)$ in (1.8). Here

$$b_n = \sum_{k=1}^{n} d_k^{(n)} c_k, \; d_k^{(n)} = \frac{1}{n!} \left\{ \frac{\partial}{\partial \eta^n} [\psi(\eta)]^k \right\}_{\eta=0} .$$

The function (1.9) is a meromorphic function with poles which are the preimages of the poles of R_λ. From this it follows that the domains D, Δ and the function $\psi(\eta)$ could be chosen so that the point $\eta^* = \psi^{-1}(\lambda^*)$ belongs to the region of convergence of the series (1.9). This gives a method for solving problems (1) and (2) mentioned above.

Let us consider a simple example. Assume that all the characteristic numbers λ_k of the integral operator are real and negative, i.e., $\lambda_k \in (-\infty, -a]$, where $a > 0$. For the function $\psi(\eta)$, it is convenient here to take the mapping of the disk $|\eta| < 1$ on the complex plane with a cut along the negative real axis from a to $-\infty$:

$$\lambda = \psi(\eta) = \frac{4a\eta}{(1-\eta)^2} \, .$$

Suppose that it is necessary to obtain the solution to (1.7) at $\lambda = \lambda^* = 1$. Then

$$R^*_\lambda \cong \sum_{n=0}^{m} b_n \eta^{*n} = \sum_{n=0}^{m} \eta^{*n} \sum_{k=1}^{n} d_k^{(n)} c_k = \sum_{n=0}^{m} c_n l_n^{(m)}$$

where

$$l_n^{(m)} = \sum_{k=n}^{m} d_n^{(k)} \eta^{*k} .$$

To construct optimal Monte Carlo algorithms based on approximate calculation of the resolvent the following statements obtained in [1.3] can be used (in what follows, the coefficients a_i are defined from the expansion

$$\psi(\eta) = a_1\eta + a_2\eta^2 + \ldots \quad) .$$

Proposition 1. Suppose that the conformal mapping $\lambda = \psi(\eta)$ has only simple poles lying on the boundary $|\eta| = 1$ and $|a_i| \le c < \infty$. Then the coefficients $l_n^{(m)}$ are uniformly bounded if $c|\eta^*| \le |\, 1 - |\eta^*|\, |$. □

Proposition 2. Suppose that the conformal mapping $\lambda = \psi(\eta)$ maps the set $|\eta| < 1$ on a convex domain. Then the coefficients $l_n^{(m)}$ are uniformly bounded if

$$\left| \frac{a_1|\eta^*|}{1-|\eta^*|} \right| \le 1 \, .$$ □

The computational cost of the corresponding Monte Carlo algorithms for some classes of mappings $\lambda = \psi(\eta)$ is estimated in [1.3], where the relation of the method of transformation of the spectral parameter to iterative processes for solving integral equations is also discussed. Monte Carlo implementation of the resolvent iteration is investigated in Chap. 5 on the basis of a vector representation.

We mention here also a numerical method for solving integral equations of the second kind based on the Fredholm representation [1.4]:

$$\varphi(x) = h(x) + \frac{1}{D} \int_X D(x, x')\varphi(x')\, dx'$$

where

$$D = \sum_{n=0}^{\infty} \frac{(-1)^n}{n!} c_n, \; D(x, x') = \sum_{n=0}^{\infty} \frac{(-1)^n}{n!} B_n(x, x').$$

Here $c_0 = 1$, $B_0(x, x') = k(x, x')$, $c_n = \int_X B_{n-1}(x, x)\, dx \;\; [n > 0]$,

$$B_n(x, x') = \int_X \dots \int_X \begin{vmatrix} k(x, x') & k(x, x_1) & \dots & k(x, x_n) \\ k(x_1, x') & k(x_1, x_1) & \dots & k(x_1, x_n) \\ \dots & \dots & \dots & \dots \\ k(x_n, x') & k(x_n, x_1) & \dots & k(x_n, x_n) \end{vmatrix} dx_1 \dots dx_n.$$

This representation holds at least for bounded X and $k(x, x')$, and is analogous to the Cramer representation for the case of a system of linear algebraic equations. In [1.5] a Monte Carlo method for calculating D and $D(x, x')$ is proposed. In this method the approximate solution of the integral equation is represented in the form of a weighted sum of solutions of a series of linear algebraic equations whose dimensions are not large.

1.2 Simple Facts from Convergence Theory for Random Functions

Consider a random function (r.f.)

$$\xi(t) = \xi(t, \omega),$$

which is a random variable for any fixed value of t. The distribution function of this random variable is defined as

$$F_1(t, x) = \mathbf{P}(\xi(t) < x).$$

Information about the random functions used in Monte Carlo methods is given in [1.1]. Note that the definition of r.f. can be given through the set of finite dimensional distribution functions

$$F_n(t_1, x_1; \dots; t_n, x_n) = \mathbf{P}(\xi(t_1) < x_1, \dots, \xi(t_n) < x_n). \tag{1.10}$$

The set (1.10) must satisfy so-called self-consistent conditions; for example, the marginal distribution functions corresponding to a given function F_n must belong to this set.

If t is a one-dimensional variable, then $\xi(t)$ is called a random function; if t is multi-dimensional, $\xi(t)$ is called a random field. Next we consider some questions of the weak convergence theory for sequences of r.f. Notice that the weak convergence of finite dimensional distributions is defined here as the convergence of $F_n^{(m)}$ to F_n as $m \to \infty$, $n = 1, 2, \dots$ at all points of discontinuity of F_n [1.6].

Assume that in the space of samples of r.f. $\xi = \xi(t)$ a metric L is introduced. Consider a set Φ of functionals $\varphi(\xi)$ which are continuous in this metric: $\varphi \in \Phi$. We shall say that the sequence $\xi_n(t)$ is weakly convergent to $\xi(t)$ in the metric L (this will be denoted by $\xi_n \Rightarrow \xi$) if the distribution function of the random variable $\varphi(\xi_n)$ weakly converges to the distribution function of $\varphi(\xi)$ for all $\varphi \in \Phi$.

Suppose now that $t \in T$ where T is a bounded closed set, and the samples of $\xi(t, \omega)$ are continuous (for almost all ω). Then in the sample space of ξ it is natural to introduce a metric as follows:

$$\rho(\xi_1, \xi_2) = \max_t |\xi_1(t) - \xi_2(t)|. \tag{1.11}$$

The following statement is in fact the known [1.6] *Kolmogorov* criterion for slow convergence of continuous r.f.

Lemma 1.3. If the finite dimensional distributions of the processes $\xi_n(t)$ weakly converge to finite dimensional distributions of $\xi(t)$, and for arbitrary t_1, t_2 and some constants c, γ, $\alpha > 0$ the following inequality holds

$$\mathbf{M}|\xi_n(t_2) - \xi_n(t_1)|^\gamma \le c\,|t_2 - t_1|^{1+\alpha} \tag{1.12}$$

then $\xi_n \Rightarrow \xi$ in the metric (1.11). □

On the basis of the *Kolmogorov* criterion in [1.7] the convergence of estimates in the dependence sampling technique was investigated; in this method the values of a function (e.g., an integral with a parameter) are calculated using the same set of samplings.

Let $f(x) = \mathbf{M}\xi(x, \omega)$ and $\varepsilon(x, \omega) = \xi(x, \omega) - f(x)$. Say that $\xi \in \Gamma_k$ if

$$\left| \frac{\partial^k \varepsilon(x, \omega)}{\partial x^k} \right| \le G_k(\omega),$$

$$\int [G_k(\omega)]^p \, dP(\omega) = c_k < \infty, \quad p \ge 2.$$

Theorem 1.1. If the function $\xi(x, \omega)$ belongs to all Γ_k $(k = 1, \ldots, l)$, then the random vector function

$$\varphi_M = \sqrt{M} \left\{ M^{-1} \sum_{i=1}^{M} \frac{\partial^k \xi(x, \omega_i)}{\partial x^k} - f^{(k)}(x) \right\}, \; k = 0, \ldots, l-1$$

weakly converges to a continuous Gaussian vector function with zero mean and the covariance matrix

$$B_{jk}(x', x'') = \mathbf{M} \left\{ [\xi^{(j)}(x', \omega) - f^{(j)}(x')][\xi^{(k)}(x'', \omega) - f^{(k)}(x'')] \right\}$$

(see [1.7]).

When random functions with discontinuous realizations are considered (i.e. the samples of these r.f. can be discontinuous with probability 1), special functional spaces D are introduced [1.6]. Let $t \in [0,1]$; then D is defined as a set of real valued functions $x(t)$, such that at an arbitrary point the left and the right limits exist. It is also assumed that

$$x(t) = x(t+0),\ x(0) = x(+0),\ x(1) = x(1-0).$$

It is known that in the uniform metric (1.11) the space D is not separable. Therefore in D a special metric which is slightly weakened compared to the uniform metric is introduced.

Denote by Λ a set of continuous, monotonically increasing on $[0,1]$, real valued functions λ such that $\lambda(0) = 0$, $\lambda(1) = 1$. The metric in D is defined by

$$\rho_D(x,y) = \inf_{\lambda \in \Lambda} [\sup_t |x(t) - y(\lambda(t))| + \sup_t |t - \lambda(t)|] \quad . \tag{1.13}$$

If $x(t)$ and $y(t)$ are continuous then

$$\rho_D(x,y) = \rho(x,y) \quad .$$

The space D in the metric ρ_D is separable [1.6].

The next proposition is the *Kolmogoroff-Chentsov* slow convergence criterion in the metric (1.13).

Lemma 1.4. If the finite dimensional distributions of the processes $\xi_n(t)$ without discontinuities of second order weakly converge to the finite dimensional distributions of the process $\xi(t)$ and for arbitrary $0 \le t_1 < t_2 < t_3 \le 1$, the following inequality holds

$$\mathbf{M}[|\xi(t_3) - \xi(t_2)|\,|\xi(t_2) - \xi(t_1)|]^\gamma \le (t_3 - t_1)^{1+\alpha} \tag{1.14}$$

for some c, γ, $\alpha > 0$, then $\xi_n \Rightarrow \xi$ with respect to the metric (1.13). □

1.3 Integral Equations of the Transfer Theory and Monte Carlo Methods

Let us first consider a single velocity transfer process which can be regarded as a homogeneous Markov chain of particle (photon) collisions in a medium. After a collision, either a scattering or an absorption may occur. This changes the velocity direction of the particle. The path (a straight line) between two successive collisions is called the free path length. The angle between successive velocity directions is called the scattering angle. The particle trajectory finishes when the particle is absorbed or when it escapes from the medium.

We introduce the following notation:

$r = (x, y, z)$	is the point of the space R
$\omega = (a, b, c)$	is the unit velocity direction, $\omega \in \Omega$
$\mu_s = (\omega, \omega')$	is the cosine of the scattering angle
$w_s(\mu, r)$	is the scattering indicatrix (or a scattering phase function)
$\sigma(r)$	is the total cross section (the extinction coefficient of the flux)
σ_s	is the scattering cross section
σ_c	is the absorption cross section
$\varphi^*(r, \omega)$	is the scattering density
$\Phi(r, \omega)$	is the particle flux (the radiation intensity).

The flux Φ is defined as follows: the quantity $\Phi(r, \omega)\, ds\, d\omega$ equals the average number of particles going through a surface element (oriented perpendicular to ω at the point r) moving in a direction from the interval $[\omega, \omega + d\omega]$. It is known that

$$\varphi^*(r, \omega) = \sigma(r)\Phi(r, \omega). \tag{1.15}$$

The distribution of the free path length l for the direction ω is specified by the density

$$f_l(t) = \sigma(r) \exp\left[-\int_0^r \sigma(r(t_1))\, dr_1 \right], \tag{1.16}$$

where $r(t) = r_0 + t\omega\ r_0$ is the point where the trajectory starts (initial point). The quantity

$$\tau(t) = \int_0^t \sigma(r(t_1))\, dt_1$$

is called optical length of the segment $[r_0, r(t)]$.

To normalize the density (1.16), we assume that the density is bounded by a convex surface such that $\sigma = \sigma_c \neq 0$ outside the convex domain. The distribution density of the cosine of the scattering angle is specified by the density $w_s(\mu, r)$. If there are several scatterings specified by cross sections $\sigma_s^{(i)}(r)$ and scattering phase functions $w_s^{(i)}(\mu, r)$, then

$$\sigma_s(r) = \sum_i \sigma_s^{(i)}(r),$$

$$w_s(\mu, r) = \sigma_s^{-1}(r) \sum_i \sigma_s^{(i)}(r) w_s^{(i)}(\mu, r) \quad .$$

The absorption and scattering probabilities at a point r are defined as

$$g(r) = \sigma_c(r)/\sigma(r), \quad q(r) = \sigma_s(r)/\sigma(r),$$

respectively. Let us now describe the characteristics of the transfer process which can be calculated by the Monte Carlo method. The average number of collisions in a domain D_i approximates the integral $\int_{D_i} \varphi^*(x)\,dx$, where $x = (r, \omega)$.

If the collisions are stored with a weight $1/\sigma(r)$, then from (1.15) we obtain that this leads to an estimation of the integral $\int_{R_i} dr \int_{\Omega}(r,\omega)\,d\omega$, where R_i is the domain of coordinates r and Ω is the space of directions ω. It is known [1.8] that this integral is equal to the average length L_i of the particle's trajectory inside the domain R_i, i.e.,

$$\int_{R_i}\int_{\Omega} \Phi(r,\omega)\,d\omega = \mathbf{M}(L_i) \quad . \tag{1.17}$$

Let us give the idea of the proof. It follows from (1.16) that the set of collisions can be regarded as a nonhomogeneous Poisson random process. Therefore the probability of a collision (and the expectation of the number of collisions in a small part of the trajectory of length dl to within negligible values of higher order) is defined as follows:

$$\mathbf{P}(dl) = \sigma(r(l))dl \tag{1.18}$$

not depending on the previous history of the trajectory.

Note that the last relation is in fact the main physical assumption which is usually made [1.9] to derive, using additional assumptions about the scattering process, the integro-differential transfer equation. In view of (1.18) the relation (1.17) is obviously true in a small volume element R_i. The generalization to arbitrary domain D_i is obtained by summing (1.17) over all small volume elements.

The relation (1.17) can be generalized as follows. Let

$$I_h = \mathbf{M} \sum_{n=0}^{N} h(r_n),$$

where r_n is the chain of collisions inside $\mathbf{R}$. Then I_h can be represented as

$$I_h = \mathbf{M} \int_0^L h(r(l))\sigma(r(l))\,dl, \tag{1.19}$$

where the integral is taken over the trajectory inside the domain under study. The expression (1.19) presents,in fact, a general path estimate [1.8]; weight modifications of this estimate will be considered in Sect. 2.2.

We now consider the known [1.1, 1.10] rejection method (method of maximal cross sections, or a delta-scattering method), which simplifies the simulations in complicated domains.

Let $w(\omega, \omega'; r)$ be the distribution of the directions ω after a scattering at the point r under the condition that the previous direction ω' was fixed.

The flux $\Phi(r,\omega)$ satisfies the transfer integro-differential equation

$$(\omega, \operatorname{grad} \Phi(r,\omega)) + \sigma(r)\Phi(r,\omega) = \int \Phi(r,\omega')\sigma_s(r)w_s(\omega,\omega';r)\,d\omega' + \Phi_0(r,\omega), \tag{1.20}$$

where $\Phi_0(r,\omega)$ is the source distribution density.

Let $\sigma(r) \leq \sigma_m$. Rewrite (1.20) in the form

$$[\sigma_m - \sigma(r)]\,\Phi(r,\omega) = \int \Phi(r,\omega')[\sigma_m - \sigma(r)]\delta(\omega' - \omega)\,d\omega' \quad .$$

This equation can be regarded as a transfer equation for a fictitious medium where σ_m is the total cross section, $\sigma_s(r)$ is the cross section for the physical transfer process with the phase function $w_s(\omega,\omega';r)$ while $\sigma_m - \sigma(r)$ is the cross section of the fictitious scattering (where the direction does not change).

Direct simulation of the process described ensures that the length between the collisions has the desired distribution. Indeed, we used equivalent transformation of (1.20); thus the flux $\Phi(r,\omega)$ is the same. On the other hand, the flux $\Phi(r,\omega)$ determines the distribution of arbitrary collisions as is seen from (1.15). It is sufficient to consider the case when $\sigma = \sigma_c$ and

$$\Phi(r,\omega) = \delta(r - r_0)\delta(\omega - \omega_0).$$

Then the density of physical collisions and the density of the free length coincide. It is clear that the cross section may vary, i.e., $\sigma = \sigma_m(r)$, and the method of maximal cross section can be applied in a part of the domain.

The collision density φ^* satisfies the integral transfer equation

$$\varphi^*(x) = \int_X k(x,x')\varphi^*(x')\,dx' + f(x), \tag{1.21}$$

or in the operator form

$$\varphi^* = K^*\varphi^* + f,$$

where f is the density of the initial collisions, and

$$k(x',x) = \frac{\sigma_s(r')w_s(\mu_s,r)\exp\{-\tau(r',r)\}\sigma(r)}{\sigma(r')2\pi|r-r'|^2}\,\delta\left(\omega - \frac{r-r'}{|r-r'|}\right). \tag{1.22}$$

Here μ is the cosine of the scattering angle, i.e.,

$$\mu = (\omega', r - r')/|r - r'| \quad .$$

The function $k(x,x')$ describes the transition density of the collision chain.

The weight modifications of the statistical simulation of the transfer process are constructed on the basis of the Neumann series representation of the solution to (1.21):

$$\varphi^* = \sum_{n=0}^{\infty} K^{*n} f. \tag{1.23}$$

This series converges if there exists n_0 such that $\|K^{*n_0}\| < 1$. Simple arguments show that $K^{*n} f \in L_1$ if $f \in L_1$. In this case it is not difficult to estimate the norm of K. Indeed, from (1.22) we get

$$\int\limits_X k(x, x')\, dx = \frac{\sigma_s(r')}{\sigma(r')} = q(r').$$

Hence

$$\|K^*\|_{L_1} \le \sup_{r'} q(r').$$

Thus $\|K^*\|_{L_1} < 1$ if $q(r') \le q_0 < 1$. It may appear in practice that $\|K^*\| = 1$ but $\|K^{*n_0}\| < 1$ for sufficiently large n_0 (that means that the upper bound of the survival probability after n_0 collisions is less than 1). Note that if the domain is bounded, then $\|K^{*2}\|_{L_1} < 1$ even if $q(r') \equiv 1$. This follows from the assumption that the domain is bounded by a convex surface such that $\sigma(r) = \sigma_c(r) > 0$ outside of it.

The Monte Carlo method is often used to calculate linear functionals of the type

$$I_h = (\varphi^*, h) = \int\limits_X \varphi^*(x) h(x)\, dx.$$

The last integral exists, in particular, if $\varphi^* \in L_1$ and $h \in L_\infty$. For example, if it is necessary to evaluate an integral of φ^* over a bounded domain D, then the choice is $h(x) = 1$ if $x \in D$ and $h(x) = 0$ if $x \notin D$. Since the entries of the transfer equation are positive, the Monte Carlo algorithms can be constructed for an arbitrary function $h \ge 0$ provided $(\varphi, h) < \infty$. If $\{x_n\}$ is the chain of physical collisions, then

$$I_h = \mathbf{M}\xi,\ \xi = \sum_{n=0}^{N} h(x_n).$$

It is possible to simulate another Markov chain by including in the transition density of this chain $r(x, x')$ the δ-function from (1.22). When the weights Q_n are calculated, these δ-functions can be omitted (Sect. 1.6). Direct evaluations show that the random estimate of the type

$$\xi = \sum_{n=0}^{N} Q_n h(x_n)$$

is unbiased [1.1].

Algorithms with $Q_n \not\equiv 1$ are called weighted methods. It is clear that in the weight methods the same Markov chain can be used to solve the transfer problems with different media. The important problem here is to construct the Markov chain so that the mean squared errors of the Monte Carlo estimates are as small as possible.

To investigate the variances of the weighted estimates for the functionals of type I_h we use the adjoint integral equation of transfer

$$\varphi(x) = \int_X k(x, x')\varphi(x')\, dx' + h(x), \tag{1.24}$$

or in the operator form

$$\varphi = K\varphi + h,$$

and the representation $I_h = (f, \varphi)$, $\varphi \in L_\infty$. This simplifies all the considerations.

In what follows, we denote by ξ_x the weighted estimate for the quantity $\varphi(x)$. For example, if the distribution density of x_0 is f, then

$$\mathbf{M}\xi_{x_0} = (f, \varphi) = I_h.$$

The mean squared estimates of I_h can be represented through the function

$$\Psi(x) = \mathbf{M}\xi_x^2.$$

This function can be uniformly minimized by an appropriate choice of the Markov chain. Therefore, almost all the results of this book are concerned with the constructions of optimal estimates ξ_x. Equation (1.24) [or (1.1)] is regarded here as the direct equation (in [1.11] it is treated as an adjoint equation). It should be noted that this approach corresponds to the direct Monte Carlo scheme used in the transfer theory. In particular, in the direct simulation technique the estimate ξ_x is constructed on the "physical trajectory" which starts with a collision at the point x.

In the transfer theory, the functions of type $\varphi(x)$ are called importance functions, since the value of $\varphi(x)$ equals I_h provided $f(x') = \delta(x' - x)$, which means that a unit point source is situated at the point x. Note that different Markov chains can be adjusted to the transfer process. We considered above the case where the state space consists of the states before the collisions. It is also possible to consider a Markov chain whose state space consists of the states after the collisions. The solution of the corresponding integral equation $\varphi_1(x)$ is an importance function for a particle starting from x and

$$\varphi_1(x) = \Phi^*(x),$$

where $\Phi^*(x)$ is the solution of the corresponding integro-differential transfer equation [1.11]. By using the well-known formula for the conditional expectations, we obtain, taking into account the physical nature of the importance function, that

$$\varphi(x) = h(x) + q(r) \int\limits_{\Omega} w(\omega, \omega'; r) \Phi^*(r, \omega') \, d\omega' . \tag{1.25}$$

Analogous relations could be obtained for the importance functions of different states: before a collision, after a scattering, before a new scattering direction, etc.

These relations could be obtained also by averaging the corresponding integral equations. For example, (1.25) can be obtained as follows. The importance function $\varphi_1(x)$ for a particle after scatterings satisfies the equation

$$\varphi_1(x) = \int\limits_X k_1(x, x') \varphi_1(x') \, dx' + h_1(x).$$

Then in new variables [after the change $x = (r, \omega) \to (r, \omega')$] we integrate both sides of this equation, multiplied by $w(\omega, \omega'; r)$, with respect to ω'. This yields (1.25).

Let us now consider systems of integral equations used in the transfer theory. It is known that some wave characteristics of the radiation can be described by Stokes vector (or polarization vector):

$$\Phi(x) = \{\varphi_1(x), \varphi_2(x), \varphi_3(x), \varphi_4(x)\},$$

where $\varphi_1(x)$ is the intensity, $\varphi_2(x)$ describes the degree of polarization, $\varphi_3(x)$ determines the degree of ellipticity and $\varphi_4(x)$ describes the orientation of the polarization plane [1.11]. A set of vector functions Φ satisfying the relations

$$\varphi_1(x) \geq 0, \quad \varphi_2^2(x) + \varphi_3^2(x) + \varphi_4^2(x) \leq \varphi_1^2(x)$$

define a cone S_t.

The vector function Φ satifies the system of integral equations

$$\varphi_i(x) = \sum_{j=1}^{4} \int k_{ij}(x') \varphi_j(x') \, dx' + h_i(x). \tag{1.26}$$

This equation is adjoint to the system of transfer equations studied in [1.11]. In Sect. 5.4 we specify the entries of the matrix $\{k_{ij}(x, x')\}$.

Write the system (1.26) in the operator form

$$\Phi = \boldsymbol{K}\Phi + H.$$

Note that the operator $\boldsymbol{K}$ has the same properties as the operator $\boldsymbol{K}^*$, i.e., $\boldsymbol{K} \in [S_t \to S_t]$. Denote by K_{11} the scalar integral operator generating by the kernel $k_{11}(x, x')$. If $\|K_{11}\| < 1$, then the Neumann series for the operator $\boldsymbol{K}$ converges in the space S_t [1.11].

Consequently, if $H \in S_t$, then the solution to (1.26) is represented as follows:

$$\Phi = \sum_{n=0}^{\infty} \boldsymbol{K}^n H, \; H \in S_t. \tag{1.27}$$

This representation does not imply that the Monte Carlo method is applicable here, since the kernel functions have alternative signs. However it appears (Sect. 5.4) that all the terms in the first component of the direct estimate are nonnegative. Therefore it is possible to average the first component termwise and this gives, in view of (1.27), $\varphi_1(x)$. Averaging other components of the Stokes vector can be carried out using the majorant property of the first component. Note also that $H^{(1)} = (1,0,0,0) \in S_t$; so we can use the estimate corresponding to H as a majorant in estimating arbitrary vectors.

The vector estimate for solving systems of transfer equations is used commonly in Monte Carlo methods. It should be noted that it is not difficult to rewrite the system in a scalar form by introducing an additional discrete coordinate which takes the values 1, 2, 3, 4 and the corresponding discrete integration measure. This equation can be used to construct scalar Monte Carlo algorithms (Sect.1.6). Scalar algorithms are also used in the multi-group systems of transfer equations. In the simplest case of the multi-group approximation,the scattering is assumed to be isotropic while the energy is changed by jump; more exactly, the number of the energy group is changed from step to step.

The matrix of the transition probabilities is defined by

$$\mathbf{P}(i \to j) = p_{ij} = \frac{\sigma_{ij}}{\sigma_i^{(s)}},$$

$$\sigma_i^{(s)} = \sum_{j=i}^{m} \sigma_{ij},$$

where σ_{ij} is the transition cross section and m is the total number of groups. It is assumed that the index of the group can only increase, i.e., the energy after a scattering cannot increase.

Under these assumptions the multi group system of equations is written in a triangular form:

$$\varphi_i(x) = \sum_{j=i}^{m} \int_X k_{ij}(x,x')\varphi_j(x,x')\,dx' + h_i(x),$$

where

$$k_{ij}(x,x') = \frac{\sigma_i^{(s)}(r)\sigma_i^{(t)}(r')\exp\{-\tau_i(r,r')\}\sigma_{ij}(r')}{4\pi\sigma_i^{(t)}(r)|r-r'|^2\,\sigma_i^{(s)}(r')}.$$

Here $\sigma_i^{(t)}$ is the total cross section for the i-th group and $\tau_i(r,r')$ is the corresponding optical length of the segment $[r,r']$.

Denote by $\boldsymbol{K}$ the matrix-integral operator of the system, and use the notation K_{ij} for the scalar integral operator generated by the kernel $k_{ij}(x,x')$. In Sect.5.5 we show that

$$\rho(\boldsymbol{K}) = \max_i \rho(K_{ii}).$$

Note that the transfer equation is often used in many nonlinear problems. For instance, a conductive radiation transfer in a layer $0 \le z \le L$ of a medium heated by an external radiation is described by a system of equations

$$c\frac{\partial T}{\partial t} = \frac{\partial}{\partial z}\left(k\frac{\partial T}{\partial z}\right) + F[I_\lambda], \quad t > 0, \quad 0 < z < L \ ; \tag{1.28}$$

$$T(0,z) = T_0(z); \quad k\frac{\partial T}{\partial z}\bigg|_{z=0,L} \ ;$$

$$F[I_\lambda] = \int_{-1}^{1} d\mu \int_{0}^{\infty} \sigma_{a\lambda}(I_\lambda - I_{\lambda b}(T))\,d\lambda\,; \tag{1.29}$$

$$\mu\frac{\partial I_\lambda}{\partial z} + \sigma_\lambda I_\lambda = \sigma_{a\lambda} I_{\lambda b}(T) + \sigma_{s\lambda}\int_{-1}^{1} \tilde{g}_\lambda(\mu',\mu) I_\lambda(z,\mu')\,d\mu', \tag{1.30}$$

$$0 < z < L, \quad -1 \le \mu \le 1, \quad 0 \le \lambda \le \infty;$$

$$I(0,\mu) = I_0(\mu), \quad \mu > 0; \quad I(L,\mu) = 0, \quad \mu < 0.$$

Here T is the temperature, λ is the wave length, c is the specific heat of the medium, k is the conductivity coefficient, $I_{\lambda b} = c_1\lambda^{-5}/(\exp\{c_2/(\lambda T)\} - 1)$ is the Plank function, I_λ is the intensity of the radiation, $\sigma_{a\lambda}$ is the absorption cross section, $\sigma_{s\lambda}$ is the scattering cross section, $\sigma_\lambda = \sigma_{a\lambda} + \sigma_{s\lambda}$, and g_λ is the scattering indicatrix averaged over the azimuth scattering angle, $\mu = \cos(\vartheta)$, where ϑ is measured from the axis z. Equation (1.30) can be transformed [1.1] into an integral equation for the collision density function $\varphi(x) = \sigma_\lambda I_\lambda(z,\mu)$:

$$\varphi(x) = f(x) + \int_X k(x,x')\varphi(x')\,dx', \ x = (z,\mu,\lambda) \in X. \tag{1.31}$$

We assume that $K{:}L_1(X) \to L_1(X)$. To represent (1.29) in the form of a linear functional, let us introduce a discrete set, namely a network ($0 = z_0 < z_1 < \ldots < z_m = L$); then the temperature is approximated by

$$\tilde{T} = \sum_{i=0}^{m} T_i(t)\psi_i(z),$$

where $\tilde{T} \in H_m([0,L])$, H_m is a finite dimensional space with a basis $\{\psi_i\}$, ψ_i are functions with finite supports. By the Galerkin method we obtain the following system of equations for the vector $T = (T_0,\ldots,T_m)$:

$$\sum_{j=0}^{m} A_{ij}(T)\frac{dT_j}{dt} = \sum_{j=0}^{m} B_{ij}(T)T_j + F_j - F_j^{(0)}, \tag{1.32}$$

$$i = 0,\ldots,m; \quad t > 0 \ ;$$

$$T_j = T_j^{(0)},\ t = 0;$$

$$F_j^{(0)} = \int_0^L dz \int_0^\infty d\lambda \int_{-1}^1 \sigma_{a\lambda} I_{\lambda b}(\tilde{T}) \psi_i(z)\, d\mu\,;$$

$$F_i = \int_0^L \psi_i(z) dz \int \frac{\sigma_{a\lambda}}{\sigma_\lambda} d\lambda \int_{-1}^1 \varphi(z, \mu, \lambda)\, d\mu,\ i = 0, \ldots, m.$$

The functions

$$h_i(x) = \psi_i(z) \frac{\sigma_{a\lambda}}{\sigma_\lambda}$$

belong to $L_\infty(X)$ and the quantities F_i can be represented as scalar products

$$F_i = (h_i, \varphi) = \int_X h_i(x) \varphi(x)\, dx\,,$$

provided $\sigma_\lambda \geq \sigma_0 > 0$. These quantities can be evaluated by the Monte Carlo method when solving the integral equation (1.31) for a fixed time network, using successive solutions of equations (1.30, 32). For details see [1.13].

1.4 Other Integral Equations Solved by Monte Carlo Methods

Let us consider a three-dimensional Dirichlet problem

$$\Delta u - cu = -g, u \mid_\Gamma = \psi \tag{1.33}$$

in a domain D with the boundary Γ, where $c = \text{const} \geq 0$. We assume that the functions g, ψ and the boundary Γ are smooth so that there exists a unique solution to (1.33) which can be represented through the Green function as follows:

$$\begin{aligned} u(P_0) = {} & \frac{d_0\sqrt{c}}{4\pi d_0^2 \sinh(d_0\sqrt{c})} \int_{S(P_0)} u(s) ds \\ & + \int_{|r-P_0|<d_0} \frac{\sinh[(d_0 - |r - P_0|\sqrt{c}]}{4\pi |r - P_0| \sinh(d_0\sqrt{c})} g(r) dr, \end{aligned} \tag{1.34}$$

where $d_0 = d(P_0)$ is the distance from the point P_0 to the boundary Γ.

The first integral in (1.34) is in fact an integral over the sphere $S(P_0)$ centered at the point P_0 and tangent to the boundary Γ. The relation (1.34) can be regarded as an integral equation of the second kind with a generalized kernel describing a uniform distribution over the sphere $S(P_0)$. Following the introduction of this kernel, the first integral becomes three-dimensional. Standard Monte Carlo algorithms could be applied to these integral equations if the singularities of the kernel are included in the transition density of the Markov chain. In our case the transition density describes the transition from the point P_0 to the surface of the sphere $S(P_0)$, i.e., we come to the walk on spheres process [1.14-15].

The relation (4.2) must be considered along with the equality

$$u(P_0) = \psi(P_0)\, P_0 \in \Gamma$$

which implies that the kernel of the equation is zero if the first argument belongs to Γ. Describe now the algorithm more formally. Suppose that the solution of the Dirichlet problem is known in the region

$$\Gamma_\varepsilon = \{P \in D \cup \Gamma : d(P) < \varepsilon\}.$$

Then the solution $u(r)$ satisfies the following integral equation

$$u(r) = \int_D k(r,r')u(r')dr' + h(r), \tag{1.35}$$

where

$$k(r,r') = \begin{cases} \frac{d\sqrt{c}}{\sinh(d\sqrt{c})}\delta_r(r'), & r \notin \Gamma_\varepsilon, \\ 0 \quad , & r \in \Gamma_\varepsilon, \end{cases}$$

$$h(r) = \begin{cases} \frac{1}{4\pi}\int_D \frac{\sinh[(d-|r'-r|)\sqrt{c}]}{|r'-r|\sinh(d\sqrt{c})} g(r')dr', & r \notin \Gamma_\varepsilon, \\ u(r) \quad , & r \in \Gamma_\varepsilon. \end{cases}$$

Here $d = d(r), \delta_r(r')$ is a generalized density, describing the uniform distribution on the sphere $S(r)$. Since $d\sqrt{c}/\sinh(d\sqrt{c}) < 1$ for $c > 0$, we obtain

$$\int_D\int_D k(r,r')k(r',r'')dr'dr'' \le \int_{D-\Gamma_\varepsilon} \delta_r(r')\left(\int_D \delta_{r'}(r'')dr''\right)dr'$$
$$= \int_{D-\Gamma_\varepsilon} \delta_r(r')dr' \le 1 - \nu(\varepsilon),$$

where $\nu(\varepsilon) = \varepsilon^2/(4d_{\max}^2)$ [1.1], p. 223). Thus,

$$\|K^2\|_{L_\infty} \le 1 - \nu(\varepsilon) < 1,$$

the Neumann series for (1.35) converges, and the Monte Carlo method for solving the equation (1.35) is applicable.

In addition, it follows from this that there exists a unique solution of the differential problem (1.35), (including the generalized solutions); therefore

$$u(P_0) = \mathbf{M}\xi, \xi = h(P_0) + \sum_{n=1}^{N} Q_n h(P_n), \tag{1.36}$$

where (P_n) is the walk on spheres process which terminates after N steps in Γ_ε. This Markov chain is also called the ε-spherical process. The weights in (1.36) are defined as follows:

$$Q_0 = 1, \quad Q_n = Q_{n-1}\frac{d_{n-1}\sqrt{c}}{sh(d_{n-1}\sqrt{c})},$$

$$d_n = d(P_n), \quad n = 1, 2, \ldots \quad .$$

We now recall that we assumed that $u(r)$ is known in Γ_ε. In fact, we can use approximate values of u in Γ_ε; for example, we can take

$$u(r) \approx \psi(r^*), \quad r \in \Gamma_\varepsilon, \quad r^* \in \Gamma, \quad |r - r^*| = d(r).$$

Then we obtain a biased estimate ξ_ε , such that $\mathbf{M}\xi_\varepsilon$ approximates $u(P_0)$ to within an error of order ε. Indeed,

$$|u - u_\varepsilon| \le |\mathbf{M}\{Q_N[u(P_N) - \psi(P_N^*)]\}| \le A\varepsilon \quad ,$$

where A is a constant, i.e., the upper bound of the derivatives of u in Γ_ε.

In applied problems (e.g., in electronic optics problems) it is necessary to calculate the derivatives of the solution to the following problem:

$$\Delta u = -g, \quad u|_\Gamma = \psi. \tag{1.37}$$

At a point $P_{0x} = (x_0 + x, y_0, z_0)$ lying inside a sphere $\{x < d(P_0) = d_0\}$ we can write the following mean value relation:

$$u(P_{0x}) = \int_{S(P_0)} p(\omega, x)u(s)ds + \int_{|r-P_0|<d_0} G_x(r, d_0)g(r)dr. \tag{1.38}$$

Here $G_x(r, d_0)$ is the Green function, and $p(\omega, x)$ is its derivative

$$G_x(r, d_0) = \frac{1}{4\pi}\left(\frac{1}{|r - P_{0x}|} - \frac{d_0}{\sqrt{x^2r^2 + d_0^4 - 2d_0xx_1}}\right) \quad ,$$

$$p(\omega, x) = \frac{d_0(d_0^2 - x^2)}{4\pi(d_0^2 + x^2 - 2d_0xa)^{3/2}} \quad ,$$

where ω is a unit vector having the direction $P_1 - P_0$, and $a = a(x, \omega)$ is the cosine of the angle between the vector ω and the axis x; $x_1 = x(r - P_0)$.

Differentiating (1.38) with respect to x, we get

$$\frac{\partial u}{\partial x}(x_0, y_0, z_0) = \int\limits_{|r-P_0|<d_0} g(r)\frac{\partial}{\partial x}G_x(r, d_0)dr \,|_{x=0}$$

$$+ \quad 4\pi\mathbf{M}_\omega \left\{ \frac{\partial p(\omega, x)}{\partial x} u(P_1(\omega)) \right\}_{x=0}$$

$$= \frac{1}{4\pi} \int\limits_{|r-P_0|<d_0} \frac{x_1(d_0^3 - |r-P_0|^3)}{d_0^3|r-P_0|^3} g(r)dr$$

$$+ \quad \mathbf{M} \left\{ \frac{3a}{d_0} \left[\sum_{i=1}^{N} \frac{1}{4\pi d_i} \int\limits_{|r-P_i|<d_i} \frac{d_i - |r-P_i|}{|r-P_i|} g(r)dr \quad + \quad \Psi(P_N^*) \right] \right\} .$$

Note that one can evaluate the solution at different points inside the sphere $S(P_0)$ simultaneously, using one and the same trajectory $P_0, P_1, \ldots$, which follows from (1.38). In this method the statistical estimates are dependent and ensure a smooth behavior of the error.

Let us now consider the following boundary value problem in a domain $G \subset R^n$:

$$\Delta^m u(x) = 0, \quad x \in G, \tag{1.39}$$

$$\Delta^k u|_\Gamma = \varphi_k, \quad k = 0, \ldots, m-1, \tag{1.40}$$

where $\Delta^0 u = u$, $\partial\Gamma$ is the boundary of Γ. The function u satisfies the relation [1.3]

$$u(x_0) = -\Gamma\left(\frac{n}{2}\right) \sum_{i=1}^{m-1} \left(\frac{r}{2}\right)^{2i} \frac{\Delta^i u(x_0)}{i!\Gamma(i+n/2)} N_{x_0,r}(u), \tag{1.41}$$

where $N_{x_0,r}(u)$ is the spherical mean of the function $u(x)$:

$$N_{x_0,r}(u) = \frac{1}{\omega_n} \int\limits_{\Omega} u(x + r\omega)d\omega.$$

From (1.39, 41) we obtain

$$\begin{aligned} &\Delta^{m-1}u(x_0) = N_{x_0,r}(\Delta^{m-1}u), \\ &\Delta^{m-2}u(x_0) = -c_1\Delta^{m-1}u(x_0) + N_{x_0,r}(\Delta^{m-2}u), \\ &\ldots\ldots \\ &u(x_0) = -c_1\Delta u(x_0) - \ldots - c_{m-1}\Delta u(x_0) + N_{x_0,r}(u), \end{aligned} \tag{1.42}$$

where

$$c_i = \frac{\Gamma(n/2)(r/2)^{2i}}{i!\Gamma(i+n/2)}.$$

From this we obtain the following system of mean value relations:

$$\Delta^{m-1}u(x_0) = N_{x_0,r}(\Delta^{m-1}u),$$
$$\Delta^{m-2}u(x_0) = N_{x_0,r}(\Delta^{m-2}u) - s_1 N_{x_0,r}(\Delta^{m-1}u),$$
$$\ldots\ldots$$
$$\Delta u(x_0) = N_{x_0,r}(\Delta u) - s_1 N_{x_0,r}(\Delta^2 u) - \ldots - s_{m-2}N_{x_0,r}(\Delta^{m-1}u),$$
$$u(x_0) = N_{x_0,r}(u) - s_1 N_{x_0,r}(\Delta u) - \ldots - s_{m-1}N_{x_0,r}(\Delta^{m-1}u),$$

where

$$s_1 = c_1, s_2 = c_2 - c_1 s_1, \ldots, s_i = c_i - \sum_{j+1}^{i-1} c_j s_{i-j}.$$

On the basis of this system, a vector walk on spheres algorithm for solving the problem (1.39, 40) can be constructed [1.3].

Consider now a boundary value problem of the type

$$Lu(x) = f(x), \qquad x \in G \subset R^n \tag{1.43}$$

$$Bu|_\Gamma = 0 \quad . \tag{1.44}$$

We seek the solution in the form

$$u(x) = \int_G u_\delta(x,y) f(y) dy, \tag{1.45}$$

where $u_\delta(x,y)$ is the Green function

$$Lu_\delta(x,y) = \delta(x-y) \quad , \quad Bu_\delta|_\Gamma = 0 \quad ,$$

and L is an elliptic operator which operates on the variable x.

Let $V(x,y)$ be the fundamental solution to (1.43), i.e.,

$$LV(x,y) = \delta(x,y) \quad , \quad x,y \in \mathbf{R}^n \quad .$$

Then we seek $u_\delta(x,y)$ in the form

$$u_\delta(x,y) = V(x,y) + W(x,y),$$

where $W(x,y)$ solves the problem

$$LW(x,y) = 0 \quad , \quad BW|_\Gamma = -BV(x,y)|_{x\in\Gamma} \quad .$$

Suppose that $\eta^{(\varepsilon)}(x,y)$ is an ε-biased estimate for $W(x)$, constructed on the ε-spherical process $x_0 = x$, $x_k = x_{k-1} + \omega_k d(x_{k-1})(k > 1)$, where $\{\omega_k\}$ is a set of independent isotropic random unit vectors, $d(x_{k-1})$ is the radius of $(k-1)$-th sphere. Then the solution to (1.43, 44) at a point x can be obtained by using

$$u(x) = \mathbf{M}_\xi \mathbf{M}_{\{x_k\}} \left\{ \frac{V(x,\xi) + \eta^{(\varepsilon)}(x,\xi)}{p(\xi)} \right\} + O(\varepsilon), \tag{1.46}$$

where $p(\xi)$ is an arbitrary density function on Γ.

Note that this algorithm gives the solution at one fixed point. However, using the symmetry of the Green function, $u_\delta(x,y) = U_\delta(y,x)$, it is possible to construct the walk on spheres algorithm which calculates the solution at a set of points simultaneously.

Let us now describe the walk on boundary algorithms for solving problems of the potential theory. In this case the spectral radius of the boundary integral equations is equal to 1, and the Monte Carlo algorithms are constructed using an analytical continuation of the Neumann series [1,3].

Consider the Laplace equation

$$\Delta u(x) \quad , \quad x \in G \subset R^3 \tag{1.47}$$

where the boundary ∂G is piece-wise smooth. It is convenient to handle two boundary value problems simultaneously, i.e., the inner Dirichlet problem

$$u(t) = \Psi_1(t), \quad t \in \partial G, \tag{1.48}$$

and the exterior Neumann problem

$$\begin{aligned} &\frac{\partial u}{\partial n} = \Psi_2(t), \quad t \in \partial G, \\ &\lim_{|x|\to\infty} u(x) = 0. \end{aligned} \tag{1.49}$$

It is known that it is possible to change the boundary ∂G with $\Gamma = \partial G - \Gamma_0$, where Γ_0 is a set of zero measure including corner points of ∂G. We seek the solution to (1.47, 48) in the form of a double layer potential with unknown density $\mu(t) \quad (t \in \Gamma)$

$$u(x) = \int_\Gamma \frac{\partial}{\partial n} \frac{1}{|x-t|} \mu(t) d\sigma(t) \quad , \tag{1.50}$$

where $\sigma(t)$ is a surface element of Γ, $n(t)$ is the interior normal to Γ at the point t. The density μ satisfies the boundary integral equation

$$\mu(t) = -\int_\Gamma k(t_1,t)\mu(t_1)d\sigma(t_1) + \frac{\Psi_1(t)}{2\pi} \quad , \tag{1.51}$$

where

$$k(t_1,t) = \frac{\cos \varphi_{t_1,t}}{2\pi |t_1 - t|^2} \quad ,$$

$\varphi_{t_1,t}$ is the angle between $n(t_1)$ and $t - t_1$.

We seek the solution to the problem (1.47, 49) in the form of a simple layer potential

$$u(x) = \int_{\Gamma} \frac{1}{|x-t|} \nu(t) d\sigma(t).$$

The density $\nu(t)$ satisfies the equation

$$\nu(t) = -\int_{\Gamma} k(t, t_1) \nu(t_1) d\sigma(t_1) + \frac{\Psi_2(t)}{2\pi} \quad . \tag{1.52}$$

Thus, the solution to Dirichlet and Neumann boundary value problems can be represented as linear functionals of the solutions to (1.43) and (1.52), respectively:

$$u(x) = 4\pi(h_x, \mu), \tag{1.53}$$

$$u(x) = (\nu, g_x), \tag{1.54}$$

where

$$h_x(t) = \frac{\cos \varphi_{t,x}}{4\pi |t-x|^2} \quad , \quad g_x(t) = \frac{1}{|x-t|} \quad .$$

However, the standard Monte Carlo technique is not applicable here, because the Neumann series for (1.51, 52) diverge.

Let

$$K\mu(t) = -\int_{\Gamma} k(t_1, t) \mu(t_1) d\sigma(t_1).$$

Rewrite (1.51, 1.52) as follows:

$$\mu_\lambda(t) = \lambda K \mu_\lambda(t) + f(t) \quad , \tag{1.55}$$

$$\nu_\lambda(t) = \lambda K^* \nu_\lambda(t) + e(t) \quad . \tag{1.56}$$

It is known that $\|K\| = \|K^*\| = 1$; therefore the solution to (1.55, 56) at $\lambda : |\lambda| < 1$ can be represented by uniform and absolute convergent series

$$\begin{aligned} \mu_\lambda(t) &= R_\lambda f(t) \quad , \\ \nu_\lambda(t) &= R_\lambda e(t) \quad , \end{aligned} \tag{1.57}$$

where $R_\lambda = I + \lambda K + \lambda^2 K^2 + \ldots$ is the resolvent operator.

It is known that all the characteristic numbers of the integral equations (1.55, 56) (i.e., the poles of the resolvent) are real and negative, and the maximal characteristic number (the simple resolvent pole) λ_1 is equal to -1. The value $\lambda = 1$ is not a characteristic number, consequently, the equations (1.50, 52) have unique solutions. However, the Neumann series diverge, since $\lambda_1 = -1$. Nevertheless, in order to find the solution, it is possible to construct a continuation of R_λ outside of the disk $|\lambda| < 1$.

The simplest method for this construction consists in multiplication of the resolvent by $(\lambda + 1)/2$. As a result, we obtain a new series which converges uniformly and absolutely in the disk $|\lambda| < |\lambda_2|$, where $\lambda_2 < 1$ is the second characteristic number of the integral equation (1.55). In this case we obtain (at $\lambda = 1$) the solution to (1.50) which takes the form

$$\mu(t) = \sum_{i=0}^{\infty} m_i(t) \quad , \tag{1.58}$$

where

$$\begin{aligned} &m_0 = \mu_0(t) = f(t) \quad , \\ &m_1(t) = \mu_1(t)/2 \quad , \\ &\ldots\ldots \\ &m_i(t) = (\mu_{i-1}(t) + \mu_i(t))/2 \quad , \quad i \geq 2, \\ &\mu_i(t) = K\mu_{i-1}(t) \quad , \quad i \geq 1. \end{aligned}$$

Another method of analytical continuation is based on the use of a conformal mapping $\lambda = \omega(\eta)$, such that (1.57) is transformed to an absolutely and uniformly convergent series. Then we obtain at $\eta = \eta_0 = \omega^{-1}(1)$

$$\mu(t) = \sum_{i=0}^{\infty} \eta_0^i M_i(t) \quad , \tag{1.59}$$

where

$$\begin{aligned} &\eta_0 = 1/2, \quad M_0(t) = \mu_0(t), \\ &M_1(t) = 2\mu_1(t), \\ &M_i(t) = KM_{i-1}(t) + M_{i-1}(t), \quad (i \geq 2) \end{aligned}$$

Note that here we used information about the spectrum of the integral operator K. Therefore, the solution to the adjoint integral equation (1.52) is also represented in (1.58, 59), where f and K must be replaced with $e(t)$ and $K^*(t)$, respectively.

Simple arguments show that if Γ is a sphere, then

$$|m_i(t)| \leq \sup_{\Gamma} |f| \left(\frac{1}{3}\right)^i \quad (i \geq 1).$$

Thus,

$$\left| \sum_{i=k+1}^{\infty} m_i(t) \right| \leq \frac{1}{2} \sup_{\Gamma} |f| \left(\frac{1}{3}\right)^k .$$

In the case where the transformation $\lambda = \eta/(1-\eta)$ is used, we obtain

$$\left| \sum_{i=k+1}^{\infty} \eta_0^i M_i(t) \right| \leq \frac{1}{2} \sup_{\Gamma} |f| \left(\frac{1}{3}\right)^k .$$

Let us now construct the Monte Carlo estimates for calculating the functionals (1.53, 54). Let $\nu(t)$ be the solution to (1.52) with the right-hand side $h_x(t)$. We construct a Markov chain $\{t_i\}$ on Γ with a transition density

$$p(t_{i-1} \to t_i) = k(t_i, t_{i-1}).$$

Note that in the case when G is convex, it implies that the vector $t_i - t_{i-1}$ is isotropic [1.3]. This chain is called an isotropic walk on boundary process. The point t_0 is sampled according to $h_x(t_0)$. This means that t_0 is the point of intersection of the isotropic vector $t_0 - x$ with the boundary. Then, taking $\nu(t)$ in (1.58) or (1.59), we can construct a Monte Carlo estimate such that

$$(\nu_i, \Psi_1) = \mathbf{M} Q_i \Psi_1(t_i),$$

where $Q_0 = 1, Q_i = (-1)^i$. Thus, the solution to (1.47, 48) is represented as follows:

$$u(x) = \mathbf{M}\left\{2\sum_{i=0}^{k} A_i Q_i \Psi_1(t_i)\right\} + r_k(x), \tag{1.60}$$

where the constants A_i and the remainder depend on the method of transformation used. For example, for the multiplication method $A_0 = \ldots = A_{k-1} = 1, \quad A_k = 0.5$ and

$$|r_k(x)| \le \sup_{\Gamma} |\Psi_1(t)| \left(\frac{1}{3}\right)^k .$$

Using (1.54) and representations of the type (1.58, 59) with a finite number of terms, we can construct Monte Carlo estimates for the exterior Neumann problem which are analogous to (1.60). The Markov chain $\{t_i\}$ is then constructed as follows: the initial point t_0 is sampled according to an arbitrary appropriate density $P_0(t_0)$, and the transition $t_{i-1} \to t_i$ is constructed according to the density $k(t_i, t_{i-1})$. Then

$$u(x) = \mathbf{M}\left\{\sum_{i=0}^{k} Q_i A_i \frac{1}{|x - t_i|}\right\} + r_k(x),$$

where $Q_0 = \Psi_2(t_0)/(2\pi p_0(t_0)), Q_i = -Q_{i-1}$. For the multiplication method we have

$$|r_k(x)| \le \sup_{\Gamma} \frac{1}{|x-t|}\frac{1}{4\pi}\int_{\Gamma} |\Psi_2| d\sigma \left(\frac{1}{3}\right)^k .$$

Note that one and the same Markov chain can be used to calculate simultaneously the estimates at a set of arbitrary points. This is evident in the case of the Neumann problem, since the Markov chain is not connected with the point x. In the case of the Dirichlet problem it is necessary to change only the initial weight. Indeed, let us suppose that it is necessary to calculate the solution at x and y. Then $u(y)$ is represented in the form (1.60), where

$$Q_0 = \frac{\cos\varphi_{t_0,y}|x-t_0|^2}{\cos\varphi_{t_0,x}|y-t_0|^2} \quad .$$

Note that (1.60) can also be used to calculate derivatives of the solution. For example, $\partial u(x)/\partial x_1 (x = (x_1, x_2, x_3))$ is also calculated from (1.60), where Q_0 must be taken as

$$Q_0 = \frac{(n_1(t_0)|x-t_0| - 3\cos\varphi_{t_0,x}(x_1 - t_{01}))}{|x-t_0|^2\cos\varphi_{t_0,x}} \quad ,$$

where $n(t_0) = (n_1(t_0), n_2(t_0), n_3(t_0))$.

For the Neumann problem

$$\frac{\partial u(x)}{\partial x_1} = \mathbf{M}\left\{\sum_{i=0}^{k} Q_i A_i \frac{-x_1 + t_{i1}}{|x-t_i|^3}\right\} + r_k(x),$$

$$t_i = (t_{i1}, t_{i2}, t_{i3}), |r_k(x)| \le \sup_i \left\{\frac{|-x_1 + t_{i1}|}{|x-t_i|^3}\right\} \frac{1}{4\pi} \int_\Gamma |\Psi_2| d\sigma \left(\frac{1}{3}\right)^k \quad .$$

The algorithms described can be generalized on the parabolic equations. However, in this case the boundary equations are Volterra type equations and the Neumann series converge. Then the standard Monte Carlo algorithms are applicable (see for details [1.3]).

Another example where the Volterra equations are used is the simulation of chemical reactions on the basis of the Kolmogorov equations see [1.16].

Let us consider a vector-valued Markov process with jumps: $X(t) = X_1(t), ..., X_m(t))$. The components of this vector describe the random number of molecules of different types in the system under study. The jumps occur at random time instances because of the chemical reactions which have n stages. These random events are called collisions (using the terminology of the transfer theory). These events are related in a Poisson flux with a given intensity $q(x)$, i.e., the quantity $q(X(t))dt$ is the probability that the reaction will occur in the time interval $(t, t+dt)$. Thus, after a collision the vector X is changed according to the reaction. This change is described by the transition matrix $Q(X, X')$.

Let $P(X;t)$ be the probability that $X(t) = X$. Evolution of this probability is described by the Kolmogorov equation, provided some general conditions are satisfied:

$$\frac{\partial \mathbf{P}(X;t)}{\partial t} = -q(X)\mathbf{P}(X;t) + \int_X \mathbf{P}(X';t)q(X')Q(X',X)dX'.$$

However, to construct Monte Carlo methods, it is more convenient to use the integral equation for the collision density (analogous to the approach used in the transfer theory (in Sect. 1.3):

$$\varphi^*(X;t) = \delta(X, X(0))q(X)\exp\{-q(X)t\} + \int_0^t \int \varphi^*(X', s)Q(X', X)\exp\{-q(X)(t-s)\}q(X)dX'ds,$$

where $\delta(X, X') = 1$ at $X = X'$ and $\delta(X, X') = 0$ at $X \neq X'$. This is a Volterra integral equation whose operator has zero spectral radius.

Monte Carlo methods are then used to calculate mean values of various functionals of the trajectories $X(t)$. For example, it is possible to calculate the mean time the process "diffuses" in a domain of the phase space and to estimate the parameters of the system when phase transitions occur. Numerical experiments [1.16] show that there is often no need to use large number of the reacting molecules. In [1.16] the following representation was used:

$$X(t) = X(0) + BY(t),$$

where the component $Y_i (i = 1, ..., n)$ of the vector Y describes the number of different stages of the reaction occurring in the interval $(0, t)$, and B is a $m \times n$–matrix which consists of the transition matrices of different stages. In [1.16] various Monte Carlo estimates for calculating integral characteristics of the process were constructed (the collision estimates, the free-length estimates of mathematical expectations). Note that the free-length estimate is applicable here since the random time between two collisions has exponential distribution.

Monte Carlo methods for solving the nonlinear Boltzmann equation are well-developed [1.1]. Simple physical interpretation of the Boltzmann equation and a lack of an effective mathematical tool to investigate the complex nonlinear equation generate a large number of approximate methods.

The first group of methods includes so-called algorithmic methods. These are constructed on the basis of physical arguments analogous to that used in the derivation of the Boltzmann equation.

The well-known method of this type is the direct simulation method due to G.A.Bird [1.17] and the method based on the Bernoulli scheme presented in [1.18]. The second group includes various iterative procedures. Numerical realizations of some iterations can be carried out by Monte Carlo methods. The most developed is the approach due to Haviland, where a successive linearization of the Boltzmann equation is used. However there is no strict justification of this method. The method based on branch Markov processes is justified and described in detail in [1.1]. We now describe the general schemes of some of methods mentioned above.

In the direct simulation method due to Bird and in the method based on the Bernoulli scheme the gas is considered to be an ensemble of N particles, and the volumc under study is approximated by a set of cells. The size of the cells is changed in each cell. The discrete time stop Δt is small compared to the mean time between two collisions. The evalution of the system is simulated in two steps.

Step 1. On a time step Δt, all the N molecules move according to the instant velocities. Here the boundary interactions and the exterior flux of new molecules must be taken into account .

Step 2. The collisions between the molecules in each cell are simulated and new velocities are calculated. When choosing a pair of molecules to simulate the collision, it is not necessary to take into account the distance between the molecules, since the size of a cell is small.

For simplicity, let us consider a one-component gas whose collision cross–section is σ. Suppose that a cell of volume V includes N particles having velocities $v_i (i = 1, ..., N)$. The collisions in Bird's method are simulated as follows:

1. A pair (i,j) is sampled according to the probability distribution

$$\mathbf{P}_{ij} = \frac{\omega_{ij}}{\lambda} \quad , \quad \lambda = \sum_{i-1}^{N-1} \sum_{j=i+1}^{N} \omega_{ij} \quad , \quad \omega ij = \frac{|v_j - v_i|\sigma}{V} \quad ;$$

2. The random value $\tau_{\nu+1} = 2/[N(N-1)\omega_{ij}]$ is added to the sum $\sum_{k=1}^{N} \tau_k$ and new velosities of the particles after velocity distribution are calculated.

In the Bernoulli scheme the collision is simulated as follows: for each pair (i,j)

1) the collision is simulated according to the probability $\mathbf{P}_{ij} = \omega_{ij}\Delta t$;

2) if collision occurs, then new velocities v_i, v_j are simulated; otherwise the velocities do not change.

The Haviland method, as mentioned above, is based on successive linearization of the Boltzmann equation. Let us consider the stationary case. The algorithm is based on the following iterative scheme:

$$w\frac{\partial}{\partial r} f_n = \int g\sigma dv \int [f_n(w')f_{n-1}(v') - f_n(w)f_{n-1}(v)] d\Omega, \tag{1.61}$$

where w', v' are the velocities after the collision, g is the relative velocity, σ is the differential scattering cross section, Ω is the angle between w and w'.

Thus, it is necessary to solve a linear system on each iterative step. These equations can be solved by the Monte Carlo method. We now consider an algorithm where the finite-difference and the Monte Carlo methods are combined. Suppose that the problem is homogeneous in space and the velocity space is divided into a set of cells. Then the following nondirect finite-difference scheme can be used:

$$\frac{f_\beta^{j+1} - f_\beta^j}{\Delta t} = -\nu_\beta^j(f) f_\beta^{j+1} + N_\beta^j(f).$$

Here the collision integral is split into two parts: $\nu_\beta^j(f)$ is the integral of the frequency collisions, and $N_\beta^j(f)$ is the integral of inverse collisions. These integrals are calculated by the Monte Carlo method in each cell β of the velocity space on the time step j. A generalization of this method to the nonhomogeneous case can be carried out on the basis of the splitting scheme [1.19].

Recently, a new approach for solving the Boltzmann equation by the Monte Carlo methods was developed. It is based on simulating the Markov chain described by

$$\begin{aligned} &\frac{d}{dt}\varphi(w,t) + g(w)\varphi(w,t) = \int \varphi(\omega,t)\mathbf{P}(w|\omega)d\omega, \\ &g(w) = \int \mathbf{P}(w|\omega)d\omega, \end{aligned} \tag{1.62}$$

where $d\varphi/dt$ is the full time derivative, $w = (w_1, ..., w_n)$, $w_i = (v_i, v'_j)$ are the phase coordinates of the i-th particle.

The density $\varphi(w,t)$ describes the evolution of a kinetic system of n particles. The function has the following representation

$$\mathbf{P}(w|\omega) = \sum_{i=1}^{n-1} \sum_{j=i+1}^{n} \mathbf{P}^{(2)}(w_i, w_j|\omega_i, \omega_j) \prod_{k\neq i,j}^{n} \delta(w_k - \omega_k),$$

where $\mathbf{P}^{(2)}(w_i, w_j|\omega_i, \omega_j)$ is uniquely defined by the effective collision cross–section.

However, exact simulation of the collision process is impossible, because two particles interact locally at a point. Therefore in simulations, one uses, instead of the physical collision, some approximate cross-section, distributed at the neighbourhood of the collision point. The equation can be solved by standard Monte Carlo methods. It is possible to obtain a Boltzmann equation for the one–particle distribution by integrating (1.62) with respect to $(n-1)$ coordinates and taking a limit by diminishing the neighbourhood of the interaction point. This can be made under some additional assumption, namely, if the detailed balance condition is satisfied. Mathematical justification of this method is given in [1.20].

1.5 Monte Carlo Methods for Calculating Integrals

Here we shall consider integrals of the type

$$I = \int_X \int_Y g(x,y)dxdy,$$

where X, Y are some finite dimensional Euclidian spaces. The function $g(x,y)$ may also include generalized functions, for example, the Dirac δ-function. In the last case the integration is carried out,in fact, in a space whose dimension is less than the dimension of the space $X \times Y$ [1.1].

Let $p(x,y)$ be a probability density in $X \times Y$ such that $p(x,y) \neq 0$ at the points of definition of $g(x,y)$. If the function $g(x,y)$ includes δ-function as a product, then it is also included in $p(x,y)$, and the δ-function is omitted in the relation $g(x,y)/p(x,y)$. Then the following representation holds:

$$I = \int_X \int_Y p(x,y) \frac{g(x,y)}{p(x,y)} dx dy = \mathbf{M} \frac{g(\xi,\eta)}{p(\xi,\eta)} = \mathbf{M}\zeta \quad , \tag{1.63}$$

where the vector (ξ, η) is distributed in $X \times Y$ with the density $p(x,y)$, and $\zeta = g(\xi,\eta)/p(\xi,\eta)$.

Using this representation, it is possible to construct a Monte Carlo estimate for calculating the integral I. The probability error of this estimate is determined by the variance $\mathbf{D}\zeta^2 = \mathbf{M}\zeta^2 - (\mathbf{M}\zeta)^2$, where

$$\mathbf{M}\zeta^2 = \int_X \int_Y \frac{g^2(x,y)}{p(x,y)} dx dy. \tag{1.64}$$

In this book we shall use often the importance sampling principle [1.21, 22].

Lemma 1.5. The quantity (1.64) is minimal and equal to

$$\mathbf{M}\zeta_0^2 = \left(\int_X \int_Y |g(x,y)| dx dy \right)^2 = I_0^2$$

iff

$$p(x,y) = |g(x,y)|/I_0.$$

Represent $p(x,y)$ by

$$p(x,y) = p_1(x) p_2(y|x),$$

where $p_1(x)$ is the marginal density of η under the condition that $\xi = x$. Then

$$\mathbf{M}\zeta^2 = \int_X \frac{dx}{p_1(x)} \int_Y \frac{g^2(x,y)}{p_2(y|x)} dy. \tag{1.65}$$

From this and from the importance sampling principle follows

Lemma 1.6. Let $p_2(y|x)$ be a given conditional density. Then $\mathbf{M}\zeta^2$ is minimal, iff

$$p_1(x) = c \left[\int_Y \frac{g^2(x,y)}{p_2(y|x)} dy \right]^{1/2} . \tag{1.66}$$

The last integral is in fact the mean square of the estimate

$$I_x = \int_Y g(x,y)dy = \mathbf{M}\frac{g(x,\eta)}{p_2(\eta|x)} \quad .$$

Thus Lemma 1.6 shows that the optimal choice of the random value ξ must be carried out with respect to a density which is proportional to the mean square of the contribution to the estimate of the integral. This rule agrees with the Bellmann principle [1.23] and will be often be used in our book.

We now assume that it is necessary to calculate the integral

$$I = \int_X \int_Y p(x,y)g(x,y)dxdy,$$

where $f(x,y)$ is a density of the vector (ξ, η). Introduce the notations

$$\zeta = g(\xi,\eta), \quad \mathbf{M}[\zeta|x] = \int_Y f_2(y|x)g(x,y)dy,$$

$$f_1(x) = \int_Y f(x,y)dy.$$

Here $f_1(x)$ is a density of ξ, $f_2(y|x)$ is the conditional density of η under the condition that $\xi = x, \mathbf{M}[\zeta|x]$ is the conditional mathematical expectation of ξ under the condition that $\xi = x$.

The formula of conditional mathematical expectation follows from the Fubini theorem. Thus, if the expectation $\mathbf{M}[\xi|x]$ can be evaluated for each value of x by some simple method, then the problem reduces to the Monte Carlo calculation of the integral

$$\int_X f_1(x)\mathbf{M}[\zeta|x]dx.$$

The variance then reduces to

$$\mathbf{DM}[\zeta|\xi] \le \mathbf{D}\zeta.$$

This follows from the known relation

$$\mathbf{D}\zeta = \mathbf{MD}[\zeta|\xi] + \mathbf{DM}[\zeta|\xi].$$

This equality shows that the variance of a conditional expectation is always less than the variance of the original random value. Sometimes, one sees several samples of η per each sample of ξ. Then the question is how many of samples η must be taken? Now we give a formulation of this method called the splitting technique.

Let ξ be a random variable with a probability density $f_1(x)$ and let $n(\xi)$ be an integer number defined by ξ $(n(\xi) \geq 1)$. The random values $\eta_1, ..., \eta_{n(\xi)}$ are independent and have the same conditional distribution density $f_2(y|x)$ under the condition that $\xi = x$. The splitting technique is based on the random estimate

$$\zeta_n = \frac{1}{n(\xi)} \sum_{k=1}^{n(\xi)} g(\xi, \eta_k).$$

Let $\eta = (\eta_1, ..., \eta_{n(\xi)})$. Using the formula of conditional mathematical expectations, we get

$$\mathbf{M}\xi_n = \mathbf{MM}[\zeta_n|\xi] = \mathbf{M}\left\{\frac{\mathbf{M}[\sum_{k=1}^{n(\xi)} g(\xi, \eta_k)|\xi]}{n(\xi)}\right\} = \mathbf{M}\left\{\frac{\mathbf{M}[\zeta|\xi]n(\xi)}{n(\xi)}\right\}$$

$$= \mathbf{MM}[\zeta|\xi] = \mathbf{M}\zeta = \int_X \int_Y f(x, y)g(x, y)dxdy,$$

i.e., ξ_n is an unbiased estimate of the integral. Now,

$$\mathbf{D}\zeta_n = \mathbf{DM}[\zeta_n|\xi] + \mathbf{MD}[\xi_n|\xi] = \mathbf{DM}[\zeta|\xi] + \mathbf{M}\{\mathbf{D}[\zeta|\xi]/n(\xi)\} \quad .$$

Let $t_1, t_2(x)$ be the average computing time needed to calculate one sample value of ξ, and η, ($\xi = x$ fixed), respectively. Then the average computing time per one sample of ζ_n is given by

$$t_n = t_1 + \mathbf{M}\xi[n(\xi)t_2(\xi)].$$

To minimize the quantity $S_n = t_n \mathbf{D}\zeta_n$, it is necessary to find the optimal value of $n(\xi)$. For simplicity, we assume that $n(\xi) = \text{const} = n$. In this case

$$t_n = t_1 + nt_2, \mathbf{D}\zeta_n = A_1 + A_2/n,$$

where $t_2 = \mathbf{M}t_2(\xi), A_1 = \mathbf{DM}[\zeta|\xi], A_2 = \mathbf{MD}[\zeta|\xi]$. The quantity

$$S_n = (t_1 + nt_2)(A_1 + A_2/n)$$

is minimal when $n = \sqrt{A_2 t_1/(a_1 t_2)}$.

The quantities t_1, t_2, A_1 and A_2 can be evaluated approximately in preliminary calculations.

1.6 Unbiasedness and Variance of Monte Carlo Methods

Let us consider an integral equation of the second kind

$$\varphi(x) = \int_X k(x, x')\varphi(x')dx' + h(x), \tag{1.67}$$

or in the operator form

$$\varphi = K\varphi + h,$$

where X is a finite-dimensional Euclidian space, $h \in L_\infty(X), K \in [L_\infty \to L_\infty]$. It is supposed that $\rho(K) < 1$ and even $\rho(K_1) < 1$, where K_1 is the integral operator with the kernel $|k(x, x')|$. Let $\{x_n\}$ be a Markov chain $(n = 0, 1, ..., N)$ with the transition density $p(x, x')$, such that

$$p(x) = 1 - \int_X p(x, x')dx' \geq 0$$

is interpreted as the probability that the trajectory terminates after the transition $x \to x'$; N is the random number of the last state of the chain. Let us introduce the random weights

$$Q_0 = 1, \quad Q_n = Q_{n-1}\frac{k(x_{n-1}, x_n)}{p(x_{n-1}, x_n)} \quad .$$

The random quantity

$$\xi_x = \sum_{n=0}^{N} Q_n h(x_n), \quad x_0 = x$$

is called a collision estimate [1.24, 25].

Lemma 1.7. If $\rho(K_1) < 1$, and $p(x, x') \neq 0$ when $k(x, x')\varphi(x') \neq 0$, then

$$\mathbf{M}\xi_x = \varphi(x).$$

Proof. When k and h are nonnegative, the proof can be obtained by termwise averaging of the sum representing the estimate ξ_x [1.1].

In the general case, we take an estimation of ξ_x by replacing k with $|k|$ and h with $|h|$. Then, taking into account that $\rho(K_1) < 1$, we obtain $|\mathbf{M}\xi_x| < c < \infty$. This ensures that averaging is possible. We now represent ξ_x in the form

$$\xi_x = h(x) + Q_1\xi_{x_1} \quad ,$$

hence

$$\mathbf{M}(\xi_x|x_1) = h(x) + Q_1\mathbf{M}\xi_{x_1}. \tag{1.68}$$

Averaging this expression over x_1, we obtain

$$\mathbf{M}\xi_x = h(x) + \int_X k(x, x_1)\mathbf{M}\xi_{x_1} dx_1,$$

i.e., $\mathbf{M}\xi_x = \varphi(x)$, since from $\rho(K_1) < 1$ it follows that $\rho(K) < 1$. □

Let K_p be the integral operator with the kernel $k^2(x, x')/p(x, x')$.

Lemma 1.8. If the assumptions of Lemma 1.7 are satisfied and $\rho(K_p) < 1$, then $\mathbf{M}\xi_x^2 < \infty$, and

$$\mathbf{M}\xi_x^2 = h(x)[2\varphi(x) - h(x)] + \int_X \frac{k^2(x, x')}{p(x, x')}\mathbf{M}\xi_{x'}^2 dx'. \tag{1.69}$$

The proof is analagous to that of Lemma 1.7 and is based on averaging the expression

$$\xi_x^2 = [h(x) + Q_1\xi_{x_1}]^2.$$

Note that the expression for $\mathbf{M}\xi_x^2$ was first obtained in [1.25].
In the following statement the condition $\rho(K_p) < 1$ is not used.

Lemma 1.9. Assume that $k(x, x') \geq 0, h(x) \geq 0$ and the assumptions of Lemma 1.7 are satisfied. Then the function $\Psi(x) = \mathbf{M}\xi_x^2$ is represented in the form of the Neumann series for (1.69):

$$\Psi = \sum_{n=0}^{\infty} K_p^n \chi,$$

where $\chi = h(2\varphi - h)$.
The proof is given in [1.1, p. 155] for the general case of k and h.

From (1.69) we obtain

$$\Psi = \sum_{n=0}^{m-1} K_p^n \chi K_p^m \Psi, \quad \Psi(x) = \mathbf{M}\xi_x^2 \quad ,$$

where

$$[K_p^m \Psi](x) = \int \ldots \int \frac{k^2(x_{m-1}, x_m)}{p(x_{m-1}, x_m)} \Psi(x_m) dx_1 \ldots dx_m. \tag{1.70}$$

The plots show that expressions which do not include the variable x_m must be omitted.

The minimal value of $\Psi(x)$ is reached, according to the importance sampling principle if the density $p(x, x')$ has the form

$$p(x,x') = c(x)|k(x,x')|\Psi^{1/2}(x'). \tag{1.71}$$

This relation, as well as Lemma 1.6, is related to the Bellmann principle and will often appear in our book. From (1.69, 71) we can derive a nonlinear equation for the optimal density $p(x,x')$. However in Chap. 3 we derive it using another approach which also allows us to obtain existence and uniqueness of the solution. Moreover, for $k, h \geq 0$ it is possible to construct, theoretically, an ideal estimate ξ_x with zero variance. This makes it possible to construct a numerical algorithm with small variance (Chap. 2).

In applied problems of transfer theory it is necessary to calculate linear functionals of the type

$$(f,\varphi) = (\varphi^*, h) = \int\limits_X \varphi^*(x)h(x)dx,$$

where $f \in L_1$, and φ^* is the solution of the adjoint equation $\varphi^* = K^*\varphi^* + f$, and f is a density of some physical source (Sect. 1.3). We recall that the equation $\varphi^* = K^*\varphi^* + f$ is treated in the transfer theory as a direct equation, and (1.67) as an adjoint equation.

Let $\pi(x)$ be a probability density such that $\pi(x) \neq 0$ when $f(x)\varphi(x) \neq 0$, and the point x_0 is distributed with $\pi(x)$. Then

$$(f,\varphi) = \mathbf{M}\xi, \quad \xi = \frac{f(x_0)}{\pi(x_0)}\xi_{x_0}.$$

Note that the expression for ξ coincides with the standard collision estimate

$$\xi = \sum_{n=0}^{N} Q_n h(x_n), \quad Q_0 = \frac{f(x_0)}{\pi(x_0)},$$

for the quantity (φ^*, h) [1.1, p. 151]. If

$$\mathbf{M}\xi_x^2 \leq c_1 < \infty, \quad f(x_0)/\pi(x_0) \leq c_2 < \infty,$$

then $\mathbf{M}\xi^2 < \infty$. If the function $\Psi(x) = \mathbf{M}\xi_x^2$ is defined, then the optimal density $\pi(x)$ is given by

$$\pi(x) = c|f(x)|\Psi^{1/2}(x).$$

Therefore, the uniform minimization of the function $\Psi(x)$ also leads to minimization of the variances of the estimates of the linear functionals (f,φ) for arbitrary $f \in L_1$.

Ealer, we used

$$\mathbf{M}\xi^2 = \mathbf{M}\left[\xi_{x_0}^2 \frac{f^2(x_0)}{\pi^2(x_0)}\right].$$

Note that the adjoint representation for the variance

$$\mathbf{D}\xi = (s, h[2\varphi - h]) - (f.\varphi)^2$$

was first given in [1.26]. Here

$$s(x) = \int_X \frac{k^2(x,x')}{p(x',x)} s(x')dx' + \frac{f^2(x)}{\pi(x)}.$$

As an example of utilization of (1.69), let us study the variance of a weighted algorithm for solving the transfer equation, based on the exponential transformation [1.1, p.174]. In this algorithm, the free path length is simulated according to a fictitious total cross-section (Sect. 1.3)

$$\sigma' = \sigma - c\cos\vartheta,$$

where σ is the cross-section of the medium under study, Θ is the angle between the path direction and a fixed axis. The weight factor is defined by

$$\frac{k(x,x')}{p(x,x')} = \frac{q\sigma}{1-p}(\sigma - c\cos\vartheta)^{-1}\exp\{-lc\cos\vartheta\} \quad ,$$

where q is the probability of survival of a particle after the collision, and p is the probability that the trajectory terminates. From this we obtain that for an infinite medium

$$\|K_p\| = \frac{q^2\sigma^2}{1-p}\int_0^\infty\int_{-1}^1 \frac{\exp\{-l(\sigma+c\mu)\}g(\mu)}{\sigma - c\mu} d\mu\, dl$$

$$= \frac{q^2\sigma^2}{1-p}\int_{-1}^1 \frac{g(\mu)d\mu}{\sigma^2 - c^2\mu^2} = 1.$$

Thus, the equation for c^* has the form

$$\frac{q^2\sigma^2}{1-p}\int_{-1}^1 \frac{g(\mu)d\mu}{\sigma^2 - c^2\mu^2} = 1. \tag{1.72}$$

If $c < c^*$, then the variance of the exponential transformation is finite (otherwise it may appear that the variance is infinite). In the case of isotropic scattering (i.e., when $g(\mu) = 1/2$) (1.72) takes the form:

$$\frac{1}{2}\frac{q^2}{1-p}\frac{\sigma}{c}\ln\frac{\sigma/c+1}{\sigma/c-1} = 1 \quad .$$

Thus if $1-p = q$, it coincides with the characteristic equation of Milne's problem (Sect. 2.4). Consequently, the use of $c = 1/L$ (L is the diffusion length, Sect. 2.4) for $p = 1-q$, $g(\mu) = 1/2$ does not ensure that the variance is finite.

We now estimate the computational cost of the walk on spheres algorithm for solving the Poisson equation (Sect. 1.4). Introduce the notations: D' is the closure of the domain D, Γ is the boundary of D; $d(P) = \min_{Q\in\Gamma}|P-Q|$; $\Gamma_\epsilon = \{P \in D' : d(P) < \epsilon\}$ is the ϵ-boundary of Γ, $S(P) = \{Q \in D' :: |Q-P| = d(P)\}$ is the sphere centered at P_0 and tangent to Γ.

We now describe the walk on spheres method. A Markov chain $\{P_n\}$ is defined as follows: the initial state is a fixed point P_0; the transition density $p(r, r') = \delta_r(r')$ is a uniform distribution on the sphere $S(r)$. The terminating probability $p(r)$ is defined as

$$p(r) = \begin{cases} 0, & r \notin \Gamma_\epsilon, \\ 1, & r \in \Gamma_\epsilon. \end{cases}$$

This Markov chain is called the walk on spheres process. It can be written as

$$P_n = P_{n-1} + \omega_n d(P_{n-1}), \quad n = 1, 2, \dots \quad ,$$

where $\{\omega_n\}$ is a sequence of independent isotropic unit vectors in R^3.

It is easy to verify that $p_1(r)$, the probability that the trajectory terminates (after reaching the ϵ-boundary Γ_ϵ) satisfies the inequality

$$p_1(r) \geq \nu(\epsilon) = \epsilon^2/(4d_{\max}^2).$$

Consequently, the mean number of steps $m(P_0, \epsilon)$ of the walk on spheres process, which determines the average computer time, is not larger than $\nu^{-1}(\epsilon)$.

To obtain a more exact result, it is necessary to evaluate the density $f(r)$ of the distribution of the mean number of spheres inside Γ_ϵ (the integral of $f(x)$ diverge as $\epsilon \to 0$). It is sufficient to define the density in the neighbourhood of a plane boundary. We denote by x the distance from a fixed point to the plane. It is not difficult to obtain that $f(x) \sim cx^{-1}$. Consequently, $m(P_0, \varepsilon) \leq c|\ln \varepsilon|$, as $\varepsilon \to 0$. Strict results for quite general domains can be obtained on the basis of the renewal theory [1.1].

Thus, the computational cost of the algorithm is proportional to

$$\mathbf{D}\xi_\varepsilon C|\ln \varepsilon|/\varepsilon^2 \quad ,$$

where ξ_ε is the ε-biased estimate (1.1, p. 235 and Sect. 1.4). So it is important to obtain a uniform boundedness of $\mathbf{D}\xi_\varepsilon$ as $\varepsilon \to 0$. This question is considered in [1.1].

Let us now consider the walk on spheres method for solving the biharmonic equation in a bounded domain $\Gamma \subset \mathbf{R}^m, \Gamma = \delta\Gamma$:

$$\Delta\Delta u = 0, \quad u|_\Gamma = f, \quad \Delta u|_\Gamma = g.$$

The corresponding system of integral equations has the form ($x_0 \in \Gamma \setminus \Gamma_\varepsilon$):

$$\Delta u(x_0) = \frac{1}{4\pi d_0^2} \int_{S(x_0)} \Delta u(s) ds,$$

$$u(x_0) = \frac{1}{4\pi d_0^2} \left[-\frac{d_0^2}{2m} \int_{S(x_0)} \Delta u(s) ds + \int_{S(x_0)} u(s) ds \right]. \tag{1.73}$$

We recall that d_0 is the distance from the point x_0 to the boundary Γ.

A solution can be obtained by the vector Monte Carlo algorithm (Chap. 5):

$$\xi_{x_0} = \sum_{n=0}^{N} Q_n H(x_n), \tag{1.74}$$

where $\{x_n\}$ is the walk on spheres process, $\{Q_n\}$ are the matrix weights;

$$Q_0 = \begin{bmatrix} 1 & 0 \\ 0 & 1 \end{bmatrix}, \quad Q_n = Q_{n-1} \begin{bmatrix} 1 & 0 \\ -\frac{d_n^2}{2m} & 1 \end{bmatrix} .$$

The vector $H(x_n)$ is equal to $(g(x_N), f(x_N))^T$ if $x_N \in \Gamma_\varepsilon$, and $H(x_n) = 0$ if $x_n \notin \Gamma_\varepsilon$. The variance of this algorithm is finite for arbitrary $\varepsilon > 0$ (Sect. 5.5). Moreover, the variance is uniformly bounded when $\varepsilon \to 0$.

Note that the estimate for $u(x)$ has the form $(m = 2)$:

$$\xi_{x_0}^{(2)} = \frac{g(Q^*)}{4} \sum_{n=0}^{N} d_n^2 + f(Q^*),$$

where Q^* is a boundary point nearest x_N.

1.7 Weighted Estimates for Bilinear Functionals

Let us consider a bilinear functional of the type

$$(\varphi_1, \varphi_2^* t) = \int_X \varphi_1(x) \varphi_2^*(x) t(x)\, dx, \tag{1.75}$$

where $t \in L_\infty$, and the functions $\varphi_1(x), \varphi_2^*(x)$ satisfy the equations

$$\varphi_1(x) = \int_X k_1(x,y) \varphi_1(y)\, dy + h(x), \tag{1.76}$$

$$\varphi_2^*(x) = \int_X k_2(x,y) \varphi_2^*(y)\, dy + f(x), \tag{1.77}$$

or in the operator form

$$\varphi_1 = K_1 \varphi_1 + h, \quad \varphi_2^* = K_2^* \varphi_2^* + f.$$

Note that (1.76) is considered in L_∞ and (1.77) in L_1. The operators K_1, K_2 are bounded in the norm of L_∞ and satisfy the conditions

$$\rho(K_1) < 1, \quad \rho(K_2) < 1.$$

In Monte Carlo methods, the functionals of the type (1.75) arise when the perturbation theory is used. For simplicity, let us consider the case ([1.27]) when $k_2(y,x)$ is a substochastic kernel, i.e.,

$$k_2(y,x) \geq 0, \qquad \int\limits_Y k_2(y,x)\,dx \leq 1,$$

and $f(x)$ is a probability density. To calculate the functionals of the type (φ_2^*, r), $r \in L_\infty$, the direct simulation can be used to solve the equation (1.77). The estimate based on the direct simulation technique has the form

$$\xi = \sum_{n=0}^{N} r(x_n),$$

where $\{x_n\}$ is a Markov chain with the transition density $p(x,x') = k_2(x,x')$ and the initial density $f(x)$. Consequently,

$$(\varphi_1, \varphi_2^* t) = \mathbf{M}\eta,$$

where

$$\eta = \sum_{n=0}^{N} t(x_n)\varphi_1(x_n). \tag{1.78}$$

Direct application of this estimate is not possible since φ_1, the solution of (1.76), is not known. However, replacing $\varphi_1(x_m)$ in (1.78) with an appropriate random estimate we obtain, according to the double randomization principle the following estimate:

$$\xi_{x_n} = h(x_n) + \sum_{m=n+1}^{N} \frac{Q_m}{Q_n} h(x_m),$$

where $\{Q_n\}$ are the weights corresponding to the equation (1.76) and the chain $\{x_n\}$:

$$Q_0 = 1, \quad Q_n = Q_{n-1}\frac{k_1(x_{n-1},x_n)}{k_2(x_{n-1},x_n)}.$$

From this we get

$$\zeta = \sum_{n=0}^{N} t(x_n) \sum_{m=n+1}^{N} \frac{Q_m}{Q_n} h(x_m). \tag{1.79}$$

This yields

$$\zeta_0 = \sum_{m=0}^{N} Q_m h(x_m) \sum_{n=0}^{m} \frac{t(x_n)}{Q_n}.$$

In [1.27] the relation

$$\mathbf{M}\zeta_0 = (\varphi_1, \varphi_2^* t)$$

is justified for (1.76) with elements having alternating signs. In Chap. 5 we will construct weighted modifications of vector estimates of the type ζ_0 . In particular, we will derive simple conditions ensuring finiteness of the variance. Applications of the estimates of type ζ_0 to the evaluation of the perturbations of the linear functionals and the group constants in the transfer theory are described in [1.27]. In [1.28], the algorithms for calculating the derivatives of the coefficient of utilization of the heat neutrons θ for a complicated reactor cell including absorbers in the form of cylinders are given. The coefficient θ is represented in the form

$$\theta = \mathbf{M}h(r_N, i_N).$$

Here r_N is the coordinate of the last collision point, i_N is the corresponding index of the collision which is considered to be an additional coordinate of the phase space. The case $i = 1$ corresponds to the capture:$h(r,i) = 1$ if $i = 1$, and $h(r,i) = 0$ if $i \neq 1$. Now we derive the algorithm for calculating the derivatives with respect to the radius and to the shift of the absorber. To do this, we use the formula of small perturbations [1.11]:

$$\delta\theta = -(\Phi\delta\sigma, \Phi^*) + (\Phi\delta\sigma_s, \Phi^{*\prime}). \tag{1.80}$$

Here Φ is the neutron flux, Φ^* is the importance function (Sect. 1.3), $\Phi^{*\prime} = \Phi(r, \omega')$, where ω' is the neutron direction after scattering at the point r. The quantity $\delta\sigma_c$ is increasing far more rapidly than $\delta\sigma_s$ when the radius or the position of the absorber is varying (e.g., 10–50 times in the case of heat neutrons). Therefore, to approximate evaluation of $\delta\theta$, it is sufficient to take only the first term in (1.80). Note that the evaluation can be improved by replacing $\delta\sigma$ with $\delta\sigma_c$, i.e.,

$$\delta\theta \approx -(\Phi\delta_c, \Phi^*). \tag{1.81}$$

The calculations made on the basis of one group correlated estimates for a typical cell show that the error of this method is less than 10 %. We now describe the algorithms for calculating the biliniar functionals of the form (1.81).

Algorithm A (Calculation of the Derivatives of θ with Respect to the Radius of the Absorber). Denote the integral in (1.81) corresponding to the increase of the radius of the absorber δr through $(\cdot)_{\delta r}$. It is known that the evaluation of the quantity $\delta r \delta\sigma_c |\mu_k|^{-1}$ can be carried out by summing the values $\delta r \delta\sigma_c |\mu_k|^{-1}$ where μ_k is the cosine of the angle between the neutron direction and the normal vector to the surface of the absorber at the point of k-th intersection. To evaluate the statistical estimate of the quantity Φ, the subsequent part of the same trajectory can be used. Thus we get the following representation

$$(\Phi\delta_c, \Phi^*)_{\delta r} = \mathbf{M}\left\{\left(\sum_k \frac{\delta r \delta\sigma_c}{|\mu_k|}\right) h(r_N, i_N)\right\},$$

where the sum is taken over all intersections of the trajectory with the surface of the absorber. To construct the estimate for the derivative, it is sufficient to devide this expression by δr.

Algorithm B (Calculation of the Derivative of θ with Respect to H, the Shift of the Absorber). Analogous arguments show that the following representation holds

$$(\Phi\delta_c, \Phi^*)_{\delta H} = \mathbf{M}\left\{\left(\sum_k \frac{\delta\sigma_c \delta H \nu_k}{|\mu_k|}\right) h(r_N, i_N)\right\},$$

where ν_k is the cosine of the angle between the direction of the shift and the normal vector to the surface of the absorber at the point of k-th intersection. Numerical results obtained by the algorithms A and B are reported in [1.28].

1.8 Calculation of the Derivatives of the Linear Functionals and the Weak Convergence of the Functional Estimates

Let us assume that all the functions in the integral equation (1.67) depend on a parameter λ, i.e.,

$$\varphi(x, \lambda) = \int_X k(x, x')\varphi(x', \lambda)\, dx' + h(x, \lambda). \tag{1.82}$$

Consider the linear functional

$$I(\lambda) = \int_X f(x, \lambda)\varphi(x, \lambda)\, dx.$$

It is assumed that $f \in L_1$, and the equation is considered in L_∞. Under some conditions, described in Sect. 1.6, the following representation holds

$$I(\lambda) = \mathbf{M}\sum_{n=0}^{N} Q_n(\lambda) h(x_n, \lambda) = \mathbf{M}\xi_x(\lambda), \tag{1.83}$$

where

$$Q_0(\lambda) = \frac{f(x_0, \lambda)}{\pi(x_0)}, \quad Q_n(\lambda) = Q_{n-1}(\lambda)\frac{k(x_{n-1}, x_n, \lambda)}{p(x_{n-1}, x_n)}.$$

Here $\{x_n\}$ is the Markov chain with the transition density $p(x, x')$ and the initial density $\pi(x)$.

Assume that we can differentiate the expression (1.83) with respect to λ under the sign of the expectation. Then

$$\frac{dI(\lambda)}{d\lambda} = \mathbf{M} \sum_{n=0}^{N} \left[\frac{d \ln Q_n(\lambda)}{d\lambda} + \frac{d \ln h(x_n, \lambda)}{d\lambda} \right] Q_n(\lambda) h(x_n, \lambda).$$

It is possible to invert the expectation and the differentiation at the point $\lambda = \lambda_0$, if there exists a function $\psi(x_n)$ such that

$$\left| \frac{d \ln Q_n(\lambda)}{d\lambda} + \frac{d \ln h(x_n, \lambda)}{d\lambda} \right| |Q_n(\lambda) h(x_n, \lambda)| \le \psi(x_n) |Q_n(\lambda_0) h(x_n, \lambda_0)|$$

for $\lambda_0 - \varepsilon < \lambda < \lambda + \varepsilon$, and

$$\mathbf{M} \sum_{n=0}^{N} |Q_N(\lambda_0) h(x_n, \lambda_0) \psi(x_n)| < \infty.$$

Note that the last condition is satisfied if $\psi(x_n) \le cn$, $\|K_1\| < 1$ (the definition of the operator K_1 is given in Sect. 1.6).

Utilizing the logarithmic derivative simplifies the construction of the algorithms for calculating the derivatives of the linear functionals of the transfer theory. Let us assume that in a zone D of the system under study the total effective cross section is given by $\sigma = \sum_{i=1}^{k} \sigma_i + \sigma_0 t$, $t \ge 0$ is a parameter, $t_0 = 1$. For simplicity we assume that the functions h and f do not depend on t, and $h(x) = 0$ for $x \in D$. In addition, at $t = t_0$ the direct simulation is used. When the free path length l is sampled, the weight must be multiplied by

$$\frac{\sigma(t,l) \exp\{-\tau(t,l)\}}{\sigma(t_0,l) \exp\{-\tau(t_0,l)\}} = \frac{\sigma(t,l)}{\sigma(t_0,l)} \exp\{-\sigma_0 \Delta l (t-1)\}.$$

Here Δl is the length of the particle trajectory inside zone D. After choosing the type of collision, the weight is multiplied by

$$\frac{\sigma_i(t,l)/\sigma(t,l)}{\sigma_i(t_0,l)/\sigma(t_0,l)} = \frac{\sigma(t_0,l)}{\sigma(t,l)} \frac{\sigma_i(t,l)}{\sigma_i(t_0,l)}.$$

Consequently, the total weight is given by

$$Q(t) = t^m \exp\{-(t-1) \Sigma\, \sigma_0 \Delta l\},$$

where the sum is taken over all runs of the particle, m is the number of collisions of the type determined by the cross section σ_0. Now, at $t = t_0 = 1$ we have

$$\frac{d \ln Q(t)}{dt} = m - \Sigma\, \sigma_0 \Delta l, \quad Q(t_0) = 1.$$

Thus at $t = t_0$

$$\frac{dI_n}{dt} = \mathbf{M} \sum_{n=0}^{N} (m - \Sigma\, \sigma_0 \Delta l)_n h(x_n).$$

This relation is obtained by differentiating under the sign of the expectation in the following representation

$$I(t) = \mathbf{M} \sum_{n=0}^{N} Q_n(t) h(x_n).$$

This differentiation is possible if t_0 lies in the interval (t_1, t_2), where

$$\left| \frac{dQ_n(t)}{dt} \right| \le q_n, \quad \mathbf{M} \sum_{n=0}^{N} q_n h(x_n) < \infty. \tag{1.84}$$

We can take $t_1 = t_0 - \varepsilon, t_2 = t_0 + \varepsilon$ and

$$q_n = [n(1+\varepsilon)^{n-1} + (\Sigma\, \sigma_0 \Delta l)(1+\varepsilon)^n] \exp\{\varepsilon(\Sigma\, \sigma_0 \Delta l)\}.$$

We now assume that D is bounded. Then $\|K\| < 1$, $\Delta l < c$. Therefore if $\|K\|(1+\varepsilon)\exp\{\varepsilon c \sigma_0\} < 1$, then

$$\mathbf{M} \sum_{n=0}^{N} q_n h(x_n) < \sum_{n=0}^{\infty} n(1+\varepsilon)^n \exp\{n\varepsilon c \sigma_0\}(K^{*n} f, h)$$
$$+ \sigma_0 c \sum_{n=0}^{\infty} n(1+\varepsilon)^n \exp\{n\varepsilon c \sigma_0\}(K^{*n} f, h) < \infty.$$

In Sect. 5.7 we shall consider arbitrary domains (not necessarily bounded). This will be done on the basis of vector representation of the sum in (1.84).We will also show that the variance of the estimate for the derivative in the transfer problems is finite almost always when the variance of the estimate of the functional under study is finite, i.e.,

$$\mathbf{D} \sum_{n=0}^{N} Q_n(\lambda) h(x_n, \lambda) < \infty.$$

Note that if σ_0 is the absorption cross section, then $m \equiv 0$ and

$$\frac{d \ln Q(t)}{dt} = -\Sigma \sigma_0 \Delta l.$$

From this it follows that

$$\frac{d \ln Q(\sigma_c)}{d\sigma_c} = \Sigma \Delta l.$$

It is not difficult to verify that this expression is valid also when $\sigma_c = 0$, i.e., when the total cross section is equal to σ_s.

We will now discuss some features of the calculation of the derivatives when the method of maximal cross section is used (Sect. 1.3). As an example, let us consider the algorithm of the calculation of the derivatives with respect to the cross sections in different zones of a system. Such algorithms can be constructed by differentiating the corresponding weights. Since the δ-scattering indicatrix is singular, it is necessary to regard the number of the i-th type of collision as an additional coordinate of the phase space. Then the collision, considered as a state of the Markov chain, is determined after the choice of the type of collision (in our case a δ-scattering, a scattering or a capture). The collision density is equal to $\varphi = \varphi(r, \omega, i)$, i.e., it is stratified according to the types of collisions. The kernel of the integral transfer equation for the function $\varphi(r, \omega, i)$ is defined as follows:

$$k(r', \omega', i' \to r, \omega, i) = \frac{\sigma_i(r') w_s(r', \mu, i') \exp\{-\tau(r, r')\} \sigma(r)}{\sigma(r') 2\pi |r - r'|^2} \delta\left(\omega - \frac{r - r'}{|r - r'|}\right). \tag{1.85}$$

Here $w_s(r', \mu, i')$ is the indicatrix of the i-th type of collision ($w_s \equiv 0$ for the absorption).

In the integral transfer equation a sum over the values of the coordinate i appears. We suppose that in the k-th zone the capture cross section $\sigma_c^{(k)}$ is changing so that the quantity $\sigma_{\max}$ (Sect. 1.3) does not change. Then in the kernel (1.85) only the quantities

$$\sigma_1^{(k)} = \sigma_{\max} - \sigma^{(k)}, \; \sigma_2^{(k)} = \sigma_c^{(k)}$$

which correspond to the δ-scattrering and the capture in the k-th zone are changing. Consequently, the compensation weight (Sect. 1.6) is multiplied after each δ–scattering in the k-th zone by

$$(\sigma_{\max} - \sigma^{(k)} - \Delta\, \sigma_c^{(k)})/(\sigma_{\max} - \sigma^{(k)})$$

and after each absorption, i.e., by the quantity

$$(\sigma_c^{(k)} + \Delta\, \sigma_c^{(k)})/\sigma_c^{(k)}.$$

Differentiation of the total weight yields

$$\frac{d\theta}{d\sigma_c^{(k)}} = \mathbf{M}\left\{\left(-\frac{n_k}{\sigma_{\max} - \sigma^{(k)}} - \frac{\delta(k_N, k)}{\sigma_c^{(k)}}\right) h(r_N, i_N)\right\}.$$

Here n_k is the number of δ-scatterings inside the k-th zone, k_N is the index of the zone where the capture occurred; $\delta(i, j)$ is the Kronecker symbol, and $h(r, i)$ is the indicator function of the active zone.

Let us now consider the problem of slow convergence of the functional estimate for $I(t)$. This problem is related to the convergence of the random function:

$$\sqrt{M}\left\{M^{-1} \sum_{i=1}^{M} \xi_x^{(i)}(\lambda) - I(\lambda)\right\}, \quad \lambda_1 \le \lambda \le \lambda_2.$$

By Theorem 1.1, this function weakly converges to a continuous Gaussian random function if the following conditions hold:

$$|\partial \xi_x(\lambda)/\partial \lambda| \le G_x, \; \mathbf{M} G_x^2 < \infty.$$

We now return to the transfer problem and assume that $\lambda = t$ is a coefficient which appears in the product $t\sigma_0$ in a zone D of the system under study and suppose that the direct simulation algorithm is applied. Let $1-\varepsilon_1 \leq t \leq 1+\varepsilon_2$. Then

$$\left|\frac{\partial \xi_x(t)}{\partial t}\right| \leq \sum_{n=0}^{N} q_n h(x_n), \tag{1.86}$$

where

$$q_n = [n(1+\varepsilon_2)^{n-1} + (\Sigma\,\sigma_0 \Delta l)(1+\varepsilon_2)^n] \exp\{\varepsilon_1(\Sigma\,\sigma_0 \Delta l)\}.$$

Thus, Theorem 1.1 is applicable here if

$$\mathbf{M}\left[\sum_{n=0}^{N} q_n h(x_n)\right]^2 < \infty. \tag{1.87}$$

We suppose that D is bounded. Then

$$q_n \leq cn(1+\varepsilon_2)^n \exp\{\varepsilon_1(\Sigma\,\sigma_0 \Delta l)\} = cnQ_n^{(0)}.$$

The weight factor $Q_n^{(0)}$ corresponds to the kernel

$$k^{(0)}(x,x') = k(x,x')(1+\varepsilon_2)^2 \exp\{\varepsilon_1(\sigma_0 \Delta l)\},$$

where $k(x,x')$ is the substochastic kernel of the transfer equation under study. The condition (1.87) is satisfied (Lemma 1.8), if $\rho(K_p^{(0)}) < 1$, where $K_p^{(0)}$ is the operator with the kernel

$$\frac{k^{(0)2}(x,x')}{k(x,x')} = k(x,x')(1+\varepsilon_2)^2 \exp\{2\varepsilon_1 \sigma_0 \Delta l\} \quad . \tag{1.88}$$

The simplest estimate thus is obtained when $\varepsilon_1 = 0$, i.e. when the simulation is carried out for the minimal value of the varying cross-section. In this case we have

$$\rho(K_p^{(0)}) \leq \rho(K)(1+\varepsilon_2)^2. \tag{1.89}$$

As mentioned above, the result can be extended, using conclusions of Sect. 5.7, to unbounded D.

Note also that the results can be extended to the case of indirect simulation when $p(x,x') \neq k(x,x')$. It is then necessary only to introduce additional weights of the form

$$Q_n = \frac{k(x_{n-1},x_n)}{p(x_{n-1},x_n)} Q_{n-1}, \quad Q_0 = 1.$$

Then an additional weight will also appear in (1.88), and the right-hand side of (1.89) will be multiplied by $\sup_{x,x'}[k(x,x')/p(x,x')]$.

For example, if we use a chain without absorptions, then

$$\sup_{x,x'}[k(x,x')/p(x,x')] = \sup_{x} q(x),$$

where $q(x)$ is the probability that the trajectory will terminate after the transition $x \rightarrow x'$. If σ_0 is the absorption cross-section, then $m \equiv 0$, and the condition (1.89) takes the form

$$\rho(K_p^{(0)}) \leq \rho(K).$$

Thus, in the case where $\varepsilon = 0$ we always have the weak convergence under study. Note that the weak convergence means that all the functionals of I continuous in the metric of the space C (Sect. 1.2) are estimated with the mean squared error of order $M^{-1/2}$. In particular, this is the case when the maximum of $I(\lambda)$ on a given bounded interval is considered.

2. Using Information About the Solution

This Chapter is devoted to the importance sampling technique which is based on information about the solution of the integral equation under study. The proof of the theorem on ideal estimates with zero variance is given. The importance functions which are optimal in the sense of mean squared criterion for solving a set of equations with different right-hand sides are constructed. We estimated the variance of the algorithm based on the approximate importance function. We also constructed an algorithm based on the utilization of asymptotic solutions of the transfer equation. These results are new and give more exact solutions compared to that reported in [2.1, 2]. In addition, presentation of the results is changed, as a main problem we consider here the solution of the integral equation which is adjoint to the transfer integral equation.

2.1 Importance Sampling Technique

Here we consider the theorem on ideal Monte Carlo estimates: ξ_x, the collision estimate (Sect. 1.6), and $\eta_x = Q_N/g(x_N)$, the absorption estimate [2.1, p. 156].

Theorem 2.1. Let $\rho(K) < 1$ and $k(x, x')$, $h(x)$ are nonnegative functions. Then a) if $p(x, x') = k(x, x')\varphi(x')/[K\varphi](x)$, then

$$\mathbf{M}\xi_x = \varphi(x), \ \mathbf{D}\xi_x = 0;$$

b) if $p(x, x') = k(x, x')\varphi(x')/\varphi(x)$,then

$$\mathbf{M}\eta_x = \varphi(x), \ \mathbf{D}\eta_x = 0.$$

Proof. We give the proof only for case a), because case b) is treated analogously. It is not difficult to verify (Sect. 1.6) that in case a) $\varphi^2 = \chi + K_p\varphi^2$, where $\chi = h(2\phi - h)$. Therefore, iteration of the equation $\psi = \chi + K_p\psi$ beginning with $\psi_0 = \varphi^2$ yields

$$\varphi^2 = \sum_{m=0}^{n-1} K_p^m \chi + K_p^n \varphi^2.$$

Now

$$[K_p^n \varphi^2](x) = \int \ldots \int k(x, x_1) \frac{\varphi(x_0) - h(x_0)}{\varphi(x_1)} k(x_1, x_2) \frac{\varphi(x_1) - h(x_1)}{\varphi(x_2)} \ldots$$

$$\ldots k(x_{n-1}, x_n) \frac{\varphi(x_{n-1}) - h(x_{n-1})}{\varphi(x_n)} \varphi^2(x_n)\, dx_1\, dx_2 \ldots dx_n \leq c\|K^n \varphi\| \to 0,$$

as $n \to \infty$.

Thus

$$\varphi^2 = \sum_{n=0}^{\infty} K_p^n \chi, \quad \chi = h(2\varphi - h),$$

and the theorem is proved (Sect. 1.6). □

This proof is perhaps the simplest one, provided the condition $\rho(K) < 1$ is satisfied. Note that in [2.1], a more general proof is given where only a convergence of the Neumann series for a fixed function $h(x)$ was assumed. This proof is based on using conditional expectations of the type:

$$\eta_{x_0}^{(n)} = \mathbf{M}(\xi_{x_0} | x_0, \ldots, x_n) = \sum_{k=0}^{n-1} Q_k h(x_k) + Q_n \varphi(x_n).$$

The construction of an ideal Monte Carlo estimate on the basis of Theorem 2.1 is called the ideal importance sampling technique. It means that the simulation of the transitions is carried out according to the importance functions of the corresponding states of the Markov chain (Sects. 1.3, 6). In realistic importance sampling algorithms, approximations to the importance functions are used [2.1, 3].

Let us now describe the importance sampling technique when solutions to a set of equations (which differ only on the right-hand sides) are calculated simultaneously on one and the same ensemble of chains. The set of equations has the form

$$\varphi_\sigma = K \varphi_\sigma + h_\sigma, \quad \sigma \in S.$$

We introduce a probability measure Λ on S. Let us consider the average variances

$$D_\Lambda^{(0)}(x) = \int D\xi_x(\sigma) \Lambda(d\sigma),$$

$$D_\Lambda^{(1)}(x) = \int D\eta_x(\sigma) \Lambda(d\sigma),$$

for the collision and the absorption estimates, respectively. We say that the algorithms are optimal if these variances are minimal.

Theorem 2.2. Suppose that $\rho(K) < 1$ and the functions $k(x, x')$ and $h_\sigma(x)$ are nonnegative for all σ. Then the variance $D_\Lambda^{(1)}(x)$ is minimal and equals

$$D_\Lambda^{(1)*}(x) = \varphi_0^2(x) - \int \varphi_\sigma^2(x)\Lambda(d\sigma),$$

if $p(x, x') = k(x, x')\varphi_0(x')/[K\varphi_0](x)$. Here φ_0 is the solution to

$$\varphi_0 = K\varphi_0 + h_0, \quad h_0 = \left[\int h_\sigma^2 \Lambda(d\sigma)\right]^{1/2}.$$

Proof. The proof is obtained by averaging $\eta_x^2(\sigma)$ with respect to the measure Λ [2.1]. □

Theorem 2.3. Suppose that $\|K\| = q < 1$ and the functions $k(x, x')$ and $h_\sigma(x)$ are nonnegative for all σ. Then the variance $D_\Lambda^{(0)}(x)$ is minimal and equals

$$D_\Lambda^{(0)*}(x) = \varphi_1^2(x) - \int \varphi_\sigma^2(x)\Lambda(d\sigma),$$

if $p(x, x') = k(x, x')\varphi_1(x')/[K\varphi_1](x)$. Here φ_1 is the solution to

$$\varphi_1 = [(K\varphi_1)^2 + \rho_1^2]^{1/2}, \quad \rho_1^2 = \int h_\sigma(2\varphi_\sigma - h_\sigma)\Lambda(d\sigma). \tag{2.1}$$

Proof. We obtain by averaging of the Neumann series for $\mathbf{M}\xi_x^2(\sigma)$ (Lemma 1.9) that the quantity $\int \mathbf{M}\xi_x^2(\sigma)\Lambda(d\sigma)$ is represented as the Neumann series for the equation

$$\Psi = \rho_1^2 + K_p\Psi. \tag{2.2}$$

Substituting $h_1 = \varphi_1 - K\varphi_1$ into the equation

$$\rho_1^2 = h_1(2\varphi_1 - h_1)$$

yields (2.1). The existence and uniqueness of the solution to (2.1) follows from Lemma 1.2. Consequently, (2.2) can be considered to be an equation for $\mathbf{M}\xi_x^2$ corresponding to $\varphi_1 = K\varphi_1 + h_1$. Thus by Theorem 2.1 the theorem is proved. □

We now derive a relation between the minimal values of $D_\Lambda^{(0)}(x)$ and $D_\Lambda^{(1)}(x)$. By the Cauchy-Schwarz-Buniakowsky inequality

$$\begin{aligned}
\int \mathbf{M}\xi_x^2(\sigma)\,\Lambda(d\sigma) &= \int \mathbf{M}\left[\sum_{n,m=0}^{N} Q_n Q_m h_\sigma(x_n) h_\sigma(x_m)\right] \Lambda(d\sigma) \\
&= \mathbf{M}\left[\sum_{n,m=0}^{N} Q_n Q_m \int h_\sigma(x_n) h_\sigma(x_m)\right] \Lambda(d\sigma) \\
&\le \sum_{n,m=0}^{\infty} Q_n Q_m h_\sigma(x_n) = \mathbf{M}\xi_0^2.
\end{aligned}$$

Consequently, we show by Theorem 2.1 that for

$$p(x, x') = k(x, x')\varphi_0(x')/[K\varphi_0](x)$$

the following inequality holds:

$$D_A^{(0)}(x) \le \varphi_0^2(x) - \int \varphi_\sigma(x)\, \Lambda(d\sigma) = D_\Lambda^{(1)*}(x).$$

Thus,

$$D_\Lambda^{(0)*}(x) \le D_\Lambda^{(1)*}(x).$$

This inequality shows that the estimate ξ_x has some advantage compared to η_x in the case when a set of equations of the type $\varphi_\sigma = K\varphi_\sigma + h_\sigma$ is solved. Indeed, the estimate ξ_x includes uniform information about the values of $h_\sigma(x)$ obtained along the trajectory $\{x_n\}$. In fact, the advantage of ξ_x compared to η_x appeared when solving atmospheric optics problems [2.2]. Some theoretical justification of the advantage of ξ_x is given in Sect. 5.7.

Theorems 2.2, 3 show that the optimal importance functions coincide with the quantities

$$\left[\int \mathbf{M}\xi_x^2(\sigma)\Lambda(d\sigma)\right]^{1/2}, \quad \left[\int \mathbf{M}\eta_x^2(\sigma)\Lambda(d\sigma)\right]^{1/2}.$$

The same is true for the Theorem 1.1, since $\varphi^2(x) = \mathbf{M}\xi_x^2$. Thus, the main optimization principle is realized here also, as in Lemma 1.6, i.e., the transition in the simulated Markov chain must be chosen according to the importance function which is equal to the mean squared value of the corresponding contribution to the estimate. This principle will be effectively used in Chap.3. As mentioned in Sect.1.6, this principle is related to Bellmann's principle [2.4].

2.2 Weighted Path Estimates in the Transfer Theory

Solution of transfer problems are solved sometimes by the use of the so- called path estimate which is based on the use of the function of the type $h(r_n, r_{n+1})$; this means that the contribution to the estimate is determined through the old and new spatial points of the collision chain [2.1, 5]. The physical chain of collisions

$$\{x_n = (r_n, \omega_n)\}_{n=0,1\ldots}, \; x_0 = x = (r, \omega)$$

is considered as an original model. We assume that the function $h(r, r')$ is chosen so that

$$I_x = \mathbf{M}\zeta_x^{(0)}, \; \zeta_x^{(0)} = \sum_{n=0}^{N} h(r_n, r_{n+1}) \quad \forall x,$$

where I_x is the quantity to be calculated (for simplicity we consider here one-velocity case).

For example, if $h(r_n, r_{n+1})$ is the length of the part of the segment $[r_n, r_{n+1}]$ which lies in the domain D, then I_x is equal to the total flux in D from a unit source at point x. The phase point where the particle starts from the physical source will be considered as an initial collision with δ-scattering in a special domain of the phase space (i.e., $x_0 \in D_0$).

The corresponding estimate of the flux yields

$$I_x = \frac{\sigma_s(r)}{\sigma(r)} \int_{\Omega} w_s(\omega\omega'; r) \Phi^*(r, \omega')\, d\omega', \tag{2.3}$$

where w_s is the scattering indicatrix, and $\Phi^*(r, \omega')$ is the "importance" of the particle starting at the point r in the direction ω', i.e., $\Phi^*(r, \omega')$ is the solution of the corresponding adjoint integro-differential transfer equation (Sect.1.3).

In this section we present weighted path estimates. The method we use gives convenient estimates and simple analysis of the variance.

Instead of the chain $\{x_n\}_{n=0,1,\ldots}$ we consider an equivalent chain $\{y_{n+1} = (x_n, x_{n+1})\}_{n=0,1,\ldots}$ in a phase space Y of particles $y = (x, x')$. The transition density of the new chain is defined as follows:

$$k_1(y, y_1) = k(x', x_1')\, \delta(x_1 - x'), \tag{2.4}$$

where $k(x', x_1')$ is the transition density of the chain $\{x_n\}$ (Sect. 1.3).

The weighted estimate of $\zeta_x^{(0)}$ can be constructed on the basis of (2.4) provided that the transition density of the simulated chain $p_1(x, y_1)$ also includes a factor $\delta(x_1 - x')$, which can be reduced in the expression $k_1(y, y_1)/p_1(y, y_1)$. Justification of this formal procedure is given in Sect. 1.6. We note, from (2.2), that the weight factors have the form

$$\frac{k_1(y_n, y_{n+1})}{p_1(y_n, y_{n+1})} = \frac{k(x_n, x_{n+1})}{p(x_n, x_{n+1})}.$$

Therefore we obtain for the path estimate constructed

$$\zeta_x = \sum_{n=1}^{N} Q_N h(y_n),\ y_n = (x_{n-1}, x_n), \tag{2.5}$$

where the weights Q_n coincide with the weights of the standard collision estimates.

The weighted collision estimate (2.5) is constructed for the kernel (2.4) in a new phase space Y. Thus Theorem 2.1 can be applied, and the importance function $\varphi_1(y)$ satisfies the equation

$$\varphi_1(y) = \int_Y k(y, y')\varphi_1(y')dy' + h(y). \tag{2.6}$$

We now represent the function $\varphi_1(y)$ through h and I_x.

It is clear that the direct collision estimate $\xi_y^{(0)}$ for (2.6) [here $y = (x, x')$] coincides with $\zeta_x^{(0)}$ provided that $r_0 = r$, $r_1 = r'$. Therefore

$$\varphi_1(y) = \mathbf{M}\xi_y^{(0)} = \mathbf{M}(\zeta_x^{(0)} | r_0 = r,\ r_1 = r') = h(r, r') + I_{x_1} \quad ,$$

and

$$x_1 = \left(r', \frac{r' - r}{|r' - r|} \right) .$$

Thus if the optical depth of the medium is not large, or the probability of particle survival after a collision is small, then the approximation $\varphi_1(y) \approx h(r, r')$ can be used. We now assume that it is necessary to calculate total particle flux at fixed subdomains of the domain under study. If we use the optimal criterion where the sum of variances of the estimates is minimized, then the trajectory length must be sampled according to the density $cl\sigma(r(l)) \exp\{-\tau(l)\}$ [2.1]. Note that the physical density has the form $\sigma(r(l)) \exp\{-\tau(l)\}$. Therefore, this density is almost optimal also according to the minimax criterion (if some additional symmetry conditions are satisfied).

Let us now consider the equation for the function $\Psi(y) = \mathbf{M}\xi_y^2$ (Sect. 1.6):

$$\Psi(y) = \chi(y) + \int_Y \frac{k_1^2(y, y_1)}{p_1(y, y_1)} \Psi(y_1) dy_1 ,$$

or in the operator form

$$\Psi = \chi + K_{p_1}^{(1)} \Psi .$$

From (2.4) we get

$$\|K_{p_1}^{(1)}\|_{L_\infty} = \sup_y \int_Y \frac{k_1^2(y, y_1)}{p_1(y, y_1)} dy_1 = \sup_{x'} \int_X \frac{k^2(x', x_1')}{p(x', x_1')} dx_1' = \|K_p\|_{L_\infty} ,$$

where K_p is the integral operator which defines the mean square of the standard weighted collision estimate for the kernel $k(x, x')$. The norms of the powers of the operator $K_{p_1}^{(1)}$ are evaluated analogously. Therefore, the following equality holds:

$$\rho(K_{p_1}^{(1)}) = \rho(K_p).$$

Thus if $\rho(K_p) < 1$, $h \in L_\infty$ (for example, in the problem of calculating the total flux the domain is bounded), then $D\zeta_x < \infty$. In Sect. 5.7 we show, on the basis of vector algorithms, that the condition $h \in L_\infty$ can be replaced by the assumption $\|K_p^{(h)}\| < \infty$, where $K_p^{(h)}$ is the integral operator with the kernel $h^2(r', r_1')k^2(x', x_1')/p(x', x_1')$.

2.3 Estimation of the Variance $D\xi_x$ for the Importance Sampling Technique

Below we shall consider transition densities of the type

$$p(x,x') = r(x,x')[1-p(x)],\ p(x) \le \delta_p\ ,$$

where

$$r(x,x') = \frac{k(x,x')g(x')}{[Kg](x)}, \tag{2.7}$$

$$g(x) = c[1+\varepsilon(x)]\varphi(x),\ |\varepsilon(x)| \le \delta. \tag{2.8}$$

This means that the function $g(x)$ is in some sense close to $\varphi(x)$. It is assumed that $\rho(K) < 1$ and $h(x) \ge 0$. If $\delta = \delta_p = 0$, then by Theorem 2.1 we get $D\xi_x = 0$. Now we obtain an estimation of $D\xi_x$ for small values of δ and δ_p. We put $c \equiv 1$ in (2.8) since this constant does not affect the ratio (2.7).

Let us first consider the case when $p(x) \equiv 0$. The function $g(x)$ can be formally defined as the solution to

$$g = Kg + h_g,\ \ h_g = g - Kg.$$

Therefore, for the chain $\{x_n\}$ with the transition density (2.7) the following relation holds:

$$D\xi_x^{(g)} = 0,$$

where $\xi_x^{(g)} = \sum_{n=0}^{\infty} Q_n h_g(x_n)$. A generalization of Theorem 2.1 to alternating h is given in [2.6] (Sect. 3.5).

Now we use the last relation to extract the principal part in the estimate ξ_x:

$$\xi_x = \xi_x^{(g)} + \xi_x^{(0)},$$

$$\xi_x^{(0)} = \sum_{n=0}^{\infty} Q_n[h(x_n) - g(x_n) + (Kg)(x_n)],$$

where $D\xi_x = D\xi_x^{(0)}$. Now,

$$h_0 = h - g + Kg = K(\varepsilon\varphi) - \epsilon\varphi.$$

Therefore,

$$|h_0(x)| \le c_0\delta, \quad |\xi_x^{(0)}| \le c_0\delta\xi_x^{(1)}, \tag{2.9}$$

where $\xi_x^{(1)} = \sum_{n=0}^{\infty} Q_n h_1(x_n),\ h_1(x) \equiv 1$.

To obtain the estimation of $D\xi_x$, it is necessary to show that the norm of the function $\Psi(x) = \mathbf{M}[\xi_x^{(1)}]^2$ is bounded. By Lemma 1.9 the norm of Ψ is represented as

$$\Psi = \sum_{n=0}^{\infty} K_r^n \chi, \qquad \chi = h_1(2\varphi_1 - h_1),$$

where $\varphi_1 = K\varphi_1 + h$, and the kernel of the operator K_r is determined by

$$\frac{k^2(x,x')}{r(x,x')} = k(x,x')\frac{[Kg](x)}{g(x')} = \frac{\varphi(x) - h(x) + [K(\epsilon\varphi)](x)}{[1+\epsilon(x')]\varphi(x')}.$$

We shift the numerators $\varphi - h + K(\epsilon\varphi)$ in the expression $K_r^n \chi$ to the left, as in Theorem 2.1. Then we obtain a product of fractions of the type

$$Q(x,x') = \frac{\varphi(x') - h(x') + [K(\epsilon\varphi)](x')}{[1+\epsilon(x')]\varphi(x')} k(x,x'),$$

where

$$|Q(x,x')| \le \frac{\varphi(x') + \delta\varphi(x')}{(1-\delta)\varphi(x')} k(x,x') = \frac{1+\delta}{1-\delta} k(x,x').$$

Consequently, if

$$\frac{1+\delta}{1-\delta}\rho(K) < 1, \tag{2.10}$$

then the Neumann series for Ψ converges. Thus $\|\Psi\| < \infty$, and we obtain from (2.9)

$$D\xi_x = D\xi_x^{(0)} \le c_1\delta^2.$$

We now suppose that $p(x_{n-1}) \equiv 0$ for $n \le m$ and $p(x_{n-1}) \le \delta_p$ for $n > m$. The quantity $D\xi_x$ is not decreasing as $p(x)$ increases (Sect. 1.6); thus we assume that $p(x_{n-1}) \equiv \delta_p$ for $n > m$. Let us denote by $\xi_x^{(p)}$ the collision estimate which corresponds to $p(x) \equiv \delta_p$ [i.e., when the transition density is $r(x,x')(1-\delta_p)$], and let K_p be the integral operator which appears in the equation for the quantity $\mathbf{M}\xi_x$.

Now

$$D\xi_x = D\mathbf{M}(\xi_x | x_0, \ldots, x_m) + \mathbf{M}D(\xi_x | x_0, \ldots, x_m). \tag{2.11}$$

It is obvious that

$$D\mathbf{M}(\xi_x | x_0, \ldots, x_m) \le c_1\delta^2. \tag{2.12}$$

From

$$\xi_x = \sum_{n=0}^{m} Q_n h(x_n) + Q_m \xi_{x_m}^{(p)} \tag{2.13}$$

we obtain

$$\mathbf{M}D(\xi_x | x_0, \ldots, x_m) \le \|K_r^m\| \|D\xi_x^{(p)}\|. \tag{2.14}$$

Thus

$$D\xi_x^{(p)} = \mathbf{M}\xi_x^{(p)2} - \varphi^2(x) \leq \mathbf{M}\xi_x^2(x) - \varphi^2(x) + \mathbf{M}\xi_x^{(p)2} - \mathbf{M}\xi_x^2 \leq c_1\delta^2 + d_p,$$

where

$$d_p = \| \sum_{n=0}^{\infty} [1 - (1-\delta_p)^n] K_p^n \chi \| \leq \| \sum_{n=0}^{\infty} n\delta_p K_p^n \chi \|.$$

Consequently, if

$$\frac{1+\delta}{(1-\delta)(1-\delta_p)} \rho(K) < 1, \tag{2.15}$$

then $d_p \leq c_2\delta_p$ and

$$D\xi_x^{(p)} \leq c_1\delta^2 + c_2\delta_p. \tag{2.16}$$

Using the estimation $\|K^m\| \leq c\rho^m(K)$, we obtain from (2.11, 12, 14, 16)

$$D\xi_x \leq c_1\delta^2 + c_2\delta_p \left[\rho(K)\frac{1+\delta}{1-\delta}\right]^m \tag{2.17}$$

provided that the condition (2.15) is satisfied.

Let us now investigate the behavior of the computational cost $S = tD\xi_x$ as $\delta \to 0$. Here t is the average computer time needed to calculate one sample of ξ_x. For $m \sim |\ln \delta^2|$ we obtain

$$S \sim \delta^2(|\ln\delta| + c\delta_p^{-1}) \sim \delta^2|\ln\delta|.$$

This shows that it is convenient to relate the quantity m logarithmically with the error of approximation of φ. Let us now describe the case when the functions with alternating signs are considered. To this end we consider the auxiliary equation

$$\varphi_1(x) = \int_X |k(x,x')|\varphi_1(x')\,dx' + |h(x)|.$$

Let K_1 be the operator of this equation, and let

$$p(x,x') = \frac{|k(x,x')|g(x')}{[K_1 g](x)},$$

$$Q_n^{(0)} = Q_{n-1}^{(0)} \frac{|k(x,x')|}{p(x,x')},$$

$$\xi_x^{(0)} = \sum_{n=0}^{N} Q_n^{(0)} |h(x_n)|,$$

$$\xi_x = \sum_{n=0}^{N} Q_n h(x_n).$$

Then

$$\mathbf{M}\xi_x = \varphi(x),\ |\xi_x| \le \xi_x^{(0)},\ \mathbf{M}\xi_x^2 \le \mathbf{M}\xi_x^{(0)2},$$

$$D\xi_x \le D\xi_x^{(0)} + \varphi_1^2(x) - \varphi^2(x).$$

To estimate the quantity $D\xi_x^{(0)}$ for $g = (1+\varepsilon)\varphi_1$, it is possible to use (2.17). Note that this algorithm can be applied if $\varphi_1(x) \sim |\varphi(x)|$.

2.4 Using the Asymptotic Solution to the One-Velocity Transfer Equation

Let us suppose that the half-space $z \ge 0$ is filled with a homogeneous medium where the particles are scattered and absorbed. The direction of a particle's motion is specified by μ and the cosine of the scattering angle is measured from the axis z. The following quantities are given: l, the mean free length; $q < 1$, the survival probability after a collision; $w(\mu, \mu')$, the scattering phase function, such that

$$w(\mu, \mu') = w(\mu', \mu),\ \int_{-1}^{1} w(\mu, \mu') d\mu' = 1.$$

The operator K is defined by the relation (Sect. 1.3):

$$[Kh](z,\mu) = \begin{cases} \frac{q}{l|\mu|} \int\limits_0^\infty \exp\left\{-\frac{z-z'}{l|\mu|}\right\} dz' \int\limits_{-1}^{1} w(\mu,\mu') h(z',\mu') d\mu', \\ \frac{q}{l} \int\limits_{-1}^{1} w(0,\mu') h(z,\mu') d\mu', \\ \frac{q}{l\mu} \int\limits_z^\infty \exp\left\{-\frac{z'-z}{l\mu}\right\} dz' \int\limits_{-1}^{1} w(\mu,\mu') h(z',\mu') d\mu', \end{cases} \tag{2.18}$$

where $z \ge 0$, $-1 \le \mu \le 1$, the first row corresponds to the case $\mu < 0$, the second – to $\mu = 0$, and the last – to the case $\mu > 0$.

Let $g(z,\mu)$be a function ($-\infty < z < \infty$) satisfying the equation $g = K_1 g$, where K_1 is defined from (2.18) (by integrating, in the case $\mu < 0$, from $-\infty$ to z). It is known [2.7] that

$$g(z,\mu) = \exp\{-z/L\} a(\mu),$$

where the diffusion length $L > l$ and the function $a(\mu)$ satisfy the equation

$$\left(\mu\frac{l}{L}+1\right)a(\mu)=q\int_{-1}^{1}w(\mu,\mu')a(\mu')d\mu'. \tag{2.19}$$

We will assume that

$$0<M_1\le a(\mu)\le M_2<\infty.$$

In particular,this condition is satisfied if the function $w(\mu,\mu')$ satisfies the same condition.

Let us now consider the problem of calculating the probability that a particle starting at a point (z,μ) will escape from the half-space $\{z\ge 0\}$. This probability is equal to $\varphi(z,\mu)$ if

$$h(z,\mu)=\begin{cases}\exp\{-z/|\mu|\}, & \mu<0\\ 0, & \mu\ge 0.\end{cases}$$

From the definition of K, K_1, g, h it follows that

$$[Kg](z,\mu)=[K_1g](z,\mu)-h(z,\mu)g(0,\mu)=g(z,\mu)-h(z,\mu)a(\mu). \tag{2.20}$$

Consequently, $Kg/g\le 1$. A comparison of (2.20) with the equation $\varphi=K\varphi+h$ yields

$$g=\alpha\varphi,\quad M_1\le\alpha(z,\mu)\le M_2.$$

From this, taking $c=2/(M_1+M_2)$ we obtain

$$cg=(1+\varepsilon)\varphi,\ |\varepsilon(z,\mu)|\le(M_2-M_1)/(M_2+M_1). \tag{2.21}$$

Note that for the isotropic scattering (i.e., when $w(\mu,\mu')=1/2$)

$$a(\mu)=(1+l\mu/L)^{-1},\quad (M_2-M_1)/(M_2+M_1)=l/L.$$

We will now obtain a special estimation of the variance $D\xi_x$ using the inequality $Kg/g\le 1$ and the relations (2.20, 21). Using the shift of the numerators, as described in Sect. 2.3, we obtain that

$$[K_r^n\chi](x)\le c_1[Kg](x)[K^nh](x)\le c\varphi(x)[K^nh](x).$$

Therefore,

$$D\xi_x\le c\varphi^2(x).$$

From (2.13) we analogously obtain

$$\mathbf{M}D(\xi_x|x_0,\ldots,x_m)\le c_1\varphi(x)q^m\quad,$$

provided $q(1-\delta_p)^{-1}<1$. Consequently,

$$D\xi_x\le\varphi(x)[c\varphi(x)+c_1q^m].$$

If we take the number of collisions without absorption as $|\ln\varphi/\ln q|$, then

$$D\xi_x \le c_0\varphi^2(x).$$

Thus, we obtain an estimation of the relative probability error which does not depend on φ; the corresponding cost of the algorithm is of order $|\ln\varphi|$. The cost of the direct simulation technique which gives the same accuracy is proportional to $1/\varphi$.

We will now discuss the implementation of the algorithm under study. The expression

$$k(z,\mu,z',\mu')g(z',\mu') = w(\mu,\mu')a(\mu')k_1(z,\mu,z')\exp\{-z'/L\}$$

shows that the free length must be simulated as in the case of exponential transformation with the parameter $1/L$ without escape (Sect. 1.6). The scattering is simulated according to the function $w(\mu,\mu')a(\mu')$; the integral of this function can be evaluated from (2.19) if the parameters $L, a(\mu)$ are estimated precisely enough. If q is close to 1, then the transport approximation can be used. This means that the parameters $L, a(\mu)$ which correspond to isotropic scattering with effective value of the mean free length

$$l_1 = \frac{l}{q(1-\nu)+1-q},$$

are used. Here ν is the mean cosine of the angle between the directions before and after scattering [2.7].

It is clear that the same arguments remain true when the probability of transmission $\varphi_1(z_0,\mu)$ through the plane layer of the medium $0 \le z \le z'$ for a particle starting from the right boundary in the direction $\mu < 0$. Indeed, the mean squared collision estimates (ξ_1 for a finite layer and ξ for the semi-infinite layer) are related as follows: $\mathbf{M}\xi_1^2 \le \mathbf{M}\xi^2$, and $\varphi_1(z,\mu)/\varphi(z,\mu) \to c$ as $z \to \infty$ [2.7]. The inequality $\mathbf{M}\xi_1^2 \le \mathbf{M}\xi^2$ is satisfied, since $\xi_1 \le \xi$ when the particle escapes at the right boundary, while the quantity $\mathbf{M}\xi_1^2$ is decreased if the escape is not simulated.

The main difficulty in the algorithm arises from the simulation of μ from the density

$$f(\mu') = cw(\mu,\mu')a(\mu'),$$

since the Neumann rejection method [2.1] is not effective here (the indicatrix has a sharp peak). If, however, the densities $w(\mu,\mu')$ and $a(\mu')$ have two maxima, then the following approximation can be used (excluding a neighbourhood of the point $\mu = -1$):

$$f(\mu') \approx p(\mu)w(\mu,\mu') + [1-p(\mu)]a(\mu'). \tag{2.22}$$

This approximation can easily be generalized to the three-dimensional system.

If the densities $w(\mu,\mu')$ and $a(\mu')$ are approximated by constant functions (excluding the maxima points) then (2.22) can be obtained from the representation:

$$w(\mu,\mu')a(\mu') \approx a(\mu_1^*)w(\mu,\mu') + w(\mu,\mu_2^*)a(\mu'), \tag{2.23}$$

where μ_1^*, μ_2^* are the maxima of the functions $w(\mu, \mu')$ and $a(\mu')$, respectively. This approximation was used in [2.2] for simulating radiative transfer through a layer of sea water; for $a(\mu)$, the transport approximation was used, and the function $p(\mu_o)$ in (2.23) was taken as

$$p(\mu) = q_1 \left[q_1 + 2w(\mu, -1) \left(1 + \frac{\mu}{\sigma_1 L} \right) \right]^{-1}, \tag{2.24}$$

where $\sigma_1 = l_1^{-1}$ is the total transport cross section (the extinction coefficient), q_1 is the transport value of the survival coefficient in the collision act; L is the diffusion length corresponding to the transport approximation [2.7]. Transport approximation works well because the scattering indicatrix for sea water has a sharp maximum in the forward scattering direction.

Note that the problem of optimization of the function $p(\mu)$ in (2.22) was solved in [2.8], using the approach described in Chap. 3. The numerical results reported in [2.8] have confirmed that the approximation (2.24) is effective (for the example considered) (Sect. 3.5).

The results of Chap. 3 allow us to obtain optimal value of L. Note that a consistent optimization of L and $p(\mu)$ (Sect. 3.5) was discussed in [2.8].

In conclusion, let us note that the results reported can be generalized to the case when the probability of an particle's escape is calculated by summing the weights of the escaping particles. For the semi-infinite layer $\{z > 0\}$, the function

$$h(z, \mu) = \left\{ \begin{matrix} 0, & z > 0, \\ 1, & z \le 0. \end{matrix} \right\}$$

is used.

We recall that a fictitious absorber with $\sigma = \sigma_c > 0$ is located outside of the layer (i.e., for $z \le 0$) (Sect. 1.3). In this domain the transfer process is not modified. Using the adjoint representation of the mean squared collision estimate (Sect. 1.6), it is not difficult to see that the asymptotics of the variances (to within a constant factor) for two variants of the function h also coincide in the case of the direct simulation method.

3. Nonlinear Theory of Optimization for Solving Integral Equations

Here we will consider nonlinear equations which determine characteristics of the Markov chain which gives the minimal variance of the Monte Carlo estimates if only specific transitions are optimized (for example, if the free length distribution is optimized). The corresponding estimations of the variances are obtained. To optimize the total transition, we constructed a nonlinear equation; a unique solution to this equation yields the importance function (if all the functions are positive). Asymptotic optimization of simulation of the free length and of the scattering angles in the transfer problems is constructed.

3.1 Formulation of the Problem

Let us consider an integral equation of the second kind

$$\varphi(x) = \int_X k(x, x')\varphi(x')dx' + h(x), \quad x, x' \in X, \tag{3.1}$$

or in the operator form

$$\varphi = K\varphi + h,$$

where X is the n-dimensional Euclidian space, $h \in L_\infty(X), K \in L_\infty$. In this chapter we mainly treat applications in the radiative transfer theory. Therefore we assume here that the kernel $k(x, x')$ is substochastic, i.e.,

$$k(x, x') \geq 0, \quad \int_X k(x, x')dx' \leq 1.$$

Moreover, we assume that

$$\int_X k(x, x')dx' \leq q_0 < 1.$$

Let us now consider a homogeneous Markov chain $x_0 = x,\ x_1, \ldots, x_N$ with a substochastic transition density $p(x \to x') = p(x, x')$. Here N is the random length of the chain terminating in the transition $x \to x'$ with a probability

$$p(x) = 1 - \int_X p(x, x')dx'.$$

If

$$p(x, x') \neq 0 \; \{x, x' : k(x, x')\varphi(x') \neq 0\}, \tag{3.2}$$

then (Sect. 1.6) $\varphi(x) = \mathbf{M}\xi_x$, where

$$\xi_x = \sum_{n=0}^{N} Q_n h(x_n), \quad Q_n = Q_{n-1}\frac{k(x_{n-1}, x_n)}{p(x_{n-1}, x_n)}, \quad Q_0 = 1,$$

and the quantity $\mathbf{M}\xi_x^2$ is represented in the form of the Neumann series (Sect. 1.6):

$$\mathbf{M}\xi_x^2 = h(x)[2\varphi(x) - h(x)] + \int_X \frac{k^2(x, x')}{p(x, x')}\mathbf{M}\xi_{x'}^2 dx'. \tag{3.3}$$

Note that this equation can be interpreted as the relation for the total mathematical expectation of the random value $\xi_x^2 = [h(x) + Q_1\xi_{x_1}]^2$. It is known (Sect. 2.1) that if $p(x, x') = k(x, x')\varphi(x')/[K\varphi](x)$, then $\mathbf{M}\xi_x^2 = \varphi^2(x)$ (i.e. $D\xi_x = 0$).

We recall that in Monte Carlo methods, (3.1) is called an adjoint equation [3.1]. Usually, one calculates the functionals of the form

$$(f, \varphi) = \int_X f(x)\varphi(x)dx,$$

where $f(x)$ is a probability density. If the initial point x_0 of the Markov chain is sampled from the density $f(x)$, then

$$(f, \varphi) = \mathbf{M}\varphi(x_0) = \mathbf{M}(\mathbf{M}\xi_{x_0}),$$

$$\mathbf{M}(\mathbf{M}\xi_{x_0}^2) = \int_X f(x)\mathbf{M}\xi_x^2 dx.$$

Thus, the minimization of the function $\mathbf{M}\xi_x^2$ is directly related to the minimization of the variances of the estimates for the linear functionals. Note that the method for decreasing the quantity $\mathbf{M}\xi_x^2$ can be directly generalized to the absorption estimates which has the form $Q_N h(x_N)/p(x_N)$. However, in practice one mainly uses the estimates constructed on the physical absorptions (or on the escapes from the medium) which can be represented through special collision estimates ξ_x using additional absorption states. Thus the scheme described below is quite general.

We suppose that the transition $x \to x'$ is sampled by a successive choice of two components of the vector $x' = (x^{(1)'}, x^{(2)'})$, i.e.,

$$x \to x' = x \to y \to x', \quad y = (x^{(1)'}, x^{(2)}).$$

According to this definition we set

$$k(x,x') = \int_X k_1(x,y)k_2(y,x')dy,$$

$$p(x,x') = \int_X p_1(x,y)p_2(y,x')dy,$$

where the generalized densities k_1, p_1 include the factor $\delta(y^{(2)} - x^{(2)})$, while the densities k_2, p_2 include the factor $\delta(x^{(1)'} - y^{(1)})$.

For example, to simulate the radiative transfer, it is convenient to interpret $x^{(1)}$ as the velocity, and $x^{(2)}$ as the spatial coordinates of the particle. Then the integrals with respect to y and x' are, in fact, integrals taken over $x^{(1)'}$ and $x^{(2)'}$.

This scheme simplifies the description which follows. In addition, it makes it possible to generalize the algorithm to the case where the kernel is randomized by random choice of an additional point y such that $x \to y$ and $y \to x'$ and by the corresponding weights.

Let

$$\int_X k_2(y,x')dx' \equiv \int_X p_2(y,x')dx' \equiv 1.$$

Then

$$\int_X k(x,y)dy = \int_X \int_X k_1(x,y)k_2(y,x')dydx' \le q_0 < 1. \tag{3.4}$$

In the present chapter we solve the following important problem: for the fixed density $p_1(x,y)$ find the density $p_2(x,y)$ such that the quantity $\mathbf{M}\xi_x^2(\forall x)$ is minimal provided that (3.2) is satisfied. We call this density a uniform best density $p_2(x,y)$. The notation $(\forall x)$ means "for almost all with respect to the Lebesgue measure".

Taking (1.4) into account, we assume that

$$\int_X \frac{k_1^2(x,y)}{p_1(x,y)}dy \le q < 1. \tag{3.5}$$

Denote the solution to (3.3) by $\mathbf{M}\xi_x^2(k_1)$ and $\mathbf{M}\xi_x^2(p_1)$ for $p(x,x') = k(x,x')$ and $p(x,x') = \int_X p_1(x,y)k_2(y,x')dy$, respectively. If $p(x,x') = k(x,x')$, then the integral operator of (3.3) coincides with the integral operator of (1.1), hence

$$\mathbf{M}\xi_x^2(k_1) \le c_0\varphi(x),$$

since $h(2\varphi - h) \le 2\|\varphi\|h$. Therefore, it is natural to assume that

$$\mathbf{M}\xi_x^2(p_1) \le c_1\varphi(x). \tag{3.6}$$

3.2 Investigation of the Master Equation

For an arbitrary density $p_2(y, x')$ satisfying the condition (3.2), we rewrite (3.3) as follows:

$$\mathbf{M}\xi_x^2 = h(x)[2\varphi(x) - h(x)] + \int \frac{k_1^2(x,y)}{p_1(x,y)} \left[\int_X \frac{k_2^2(y,x')}{p_2(y,x')} \mathbf{M}\xi_{x'}^2 \, dx' \right] dy. \quad (3.7)$$

Let us fix $\mathbf{M}\xi_{x'}^2$ in the inner integral and vary the transition density of the first transition $y \to x'$. According to the importance sampling principle, this integral is minimal and equal to

$$\left[\int_X k_2(y,x')(\mathbf{M}\xi_{x'}^2)^{1/2} dx' \right]^2$$

if

$$p_2(y,x') = k_2(y,x')(\mathbf{M}\xi_{x'}^2)^{1/2} \left[\int_X k_2(y,x')(\mathbf{M}\xi_{x'}^2)^{1/2} dx' \right]^{-1}.$$

Consequently,

$$\begin{aligned} g_1(x) &= h(x)[2\varphi(x) - h(x)] \\ &\quad + \int_X \frac{k_1^2(x,y)}{p_1(x,y)} \left[\int_X k_2(y,x')(\mathbf{M}\xi_{x'}^2)^{1/2} dx' \right]^2 dy \le \\ &\quad \le g_0(x) = \mathbf{M}\xi_x^2. \end{aligned} \quad (3.8)$$

Note that (3.2) is not violated, since

$$\mathbf{M}\xi_x^2 \ge (\mathbf{M}\xi_x)^2 = \varphi^2(x).$$

Rewrite (3.8) in the operator form $g_1 = G(g_0) \le g_0$. Replacing g_0 with g_1 yields

$$g_2 = G(G(g_0)) \le g_1 \le g_0.$$

Note that the essence of this kind of transformation is the optimization of the density of distribution of the second transition $y \to x'$ and repeated optimization of the first transition.

Now,

$$g_n(x) = [G^n(g_0)](x) \downarrow g(x), \quad [\forall x]$$

where $g(x) \ge \varphi^2(x)$ and

$$g(x) = h(x)[2\varphi(x) - h(x)] + \int_X \frac{k_1^2(x,y)}{p_1(x,y)} \left[\int_X k_2(y,x')g^{1/2}(x')dx' \right]^2 dy. \tag{3.9}$$

The statements which follow show that this equation determines the uniformly best modification of the simulation of the transition $y \to x'$ provided $\mathbf{M}\xi_x^2 \in L_\infty$.

Lemma 3.1. If $g \in L_\infty$ is the solution to (3.9), then for the Markov chain with

$$p_2(y,x') = k_2(y,x')g^{1/2}(x')dx' \left[\int_X k_2(y,x')g^{1/2}(x')dx' \right]^{-1} \tag{3.10}$$

we get $\mathbf{M}\xi_x^2 = g(x)$.

Proof. Denote the integral operator of (3.7) with K_g, i.e.,

$$[K_g f](x) = \int_X \frac{k_1^2(x,y)}{p_1(x,y)} \left[\int_X k_2(y,x')g^{1/2}(x')dx' \right] \frac{k_2(y,x')}{g^{1/2}(x')} f(x')dx'.$$

Then from (3.9) it follows that $g = h(2\varphi - h) + K_g g$. Iterating the last equation beginning from $g_0 = g$ yields

$$g = \sum_{n=0}^{\infty} K_g^n [h(2\varphi - h)] + \lim_{n\to\infty} K_g^n g.$$

Let us consider the function $K_g^2 g$:

$$\begin{aligned}[K_g^2 g](x) = \int\int\int\int & \frac{k_1^2(x,y)}{p_1(x,y)} \left[\int k_2(y,x')g^{1/2}(x')dx' \right] \\ & \times \frac{k_2(y,x')}{g^{1/2}(x')} \frac{k_1^2(x',y_1)}{p_1(x',y_1)} \left[\int k_2(y_1,x'')g^{1/2}(x'')dx'' \right] \\ & \times \frac{k_2(y_1,x'')}{g^{1/2}(x'')} g(x'')dy\, dx' dy_1 dx''.\end{aligned}$$

From (3.9) it follows that

$$\int \frac{k_1^2(x,y)}{p_1(x,y)} \int k_2(y,x')g^{1/2}(x')dydx' \le q^{1/2} g^{1/2}(x). \tag{3.11}$$

Therefore $K_g^2 g \le \|g\| q q^{1/2}$. It is not difficult to also obtain $K_g^n g \le \|g\| q q^{(n-1)/2}$.

So the function g is represented as a Neumann series for the equation (3.3) provided that (3.10) is true and thus, according to the statement formulated before (1.3), the lemma is proved. □

Lemma 3.2. If $g \in L_\infty$ is the solution to (3.9), then $\varphi^2 \le g \le c_1\varphi$.

Proof. The operator G_1, defined as

$$[G_1(f)](x) = \left\{ h(x)[2\varphi(x) - h(x)] + \left[\int\int k_1(x,y)k_2(y,x')f(x')dy\, dx' \right]^2 \right\}^{1/2},$$

has the following property:

$$[h(2\varphi - h) + (Kf_1)^2]^{1/2} - [h(2\varphi - h) + (Kf_2)^2]^{1/2} \le |Kf_1 - Kf_2|.$$

It is not difficult to see that $G_1(\varphi) = \varphi$.

Minimizing (3.9) by an appropriate choice of $p_1(x,y)$ yields $G_1(g^{1/2}) \le g^{1/2}$. Therefore (Lemma 1.1)

$$\varphi = \lim_{n\to\infty} G_1^n(g^{1/2}) \le g^{1/2}$$

and $\varphi^2 \le g$. The operator of (3.9) is dominated by the operator of (3.7) with $p_2 \equiv k_2$, therefore $g(x) \le \mathbf{M}\xi_x^2(p_1) \le c_1\varphi(x)$ due to Lemma 1.1 and (3.6). □

Lemma 3.3. The solution of (3.9) is unique in L_∞.

Proof. We assume that there exist two different solutions g_1, $g_2 \in L_\infty$. In the Markov chain corresponding to (3.10) with $g = g_1$, we replace consecutively the transition densities of the transitions $y \to x$ with (3.10) for $g = g_2$. Then the densities are not optimal and the quantity $\mathbf{M}\xi_x^2$ is increasing. Using consecutively the formula of total mathematical expectation of

$$\xi_{x_n}^2 = [h(x_n) + (Q_{n+1}/Q_n)\xi_{n+1}]^2, \quad n = 0,1,\ldots$$

we obtain

$$g_1 \le \sum_{n=0}^{\infty} K_{g_2}^n\,(h(2\varphi - h]) + \lim_{n\to\infty} K_{g_2}^n g_1.$$

By Lemma 3.2 we get $g_1/g_2^{1/2} \le c_1$. Consequently, we obtain by arguments analogous to that of Lemma 3.1

$$K_{g_2}^n g_1 \le c_1\|g_2\|^{1/2} qq^{(n-1)/2}.$$

Thus $g_1 \le g_2$ and vice versa. □

Theorem 3.1. There exists a unique solution g to (3.9) in L_∞. The corresponding density (3.10) gives a uniformly minimal value of $\mathbf{M}\xi_x^2 \in L_\infty$.

Proof. The existence follows from the fact that the function

$$g(x) = \lim G^n(\mathbf{M}\xi_x^2(p_1))$$

satisfies (3.9) and the uniqueness follows from Lemma 3.3. The density (3.10) is uniformly optimal since $G^n(\mathbf{M}\xi_x^2) \downarrow g(x)$ for arbitrary $\mathbf{M}\xi_x^2 \in L_\infty$. □

It is not difficult to obtain that the density (3.10) is uniformly best if there is no solution to (3.9) lying outside of L_∞. On the basis of the Neumann series for (3.7) it is possible to show that if $(\varphi - h)/\varphi \le r < 1$, then $\mathbf{M}\xi_x^2 \in L_\infty$ for the density with $g \equiv \varphi$ when $p_1 \equiv k_1$. Consequently, this scheme is improved by Theorem 3.1.

From this theorem it follows that the optimization of the total transition $x \to x'$ is related to the equation

$$g(x) = h(x)[2\varphi(x) - h(x)] + \left[\int |k(x,x')| g^{1/2}(x')\, dx'\right]^2 , \tag{3.12}$$

since $x \to x' = x \to x \to x'$. It is not difficult to verify that $\varphi^2(x)$ satisfies this equation; from this we conclude that the Markov chain for which $D\xi_x^2 = 0$, $\mathbf{M}\xi_x = \varphi(x)$ is unique. Note that this result was obtained in some sense artificially in [3.2].

If $h(x) \ne 0$ only when $\int k(x,x')dx' = 0$, then (3.12) coincides with (1.1). In the transfer theory this is a case when the probability of the escape of a particle from the system is calculated by the direct simulation technique. Note that in this case Theorem 3.1 gives a new result: if the absorption is simulated at the collision point with a probability $P(x)$, then the optimal modification of the transition $x \to x'$ after the collision is determined from (3.1) where the physical survival probability $q(x)$ is replaced with $q(x)[1 - P(x)]^{-1/2}$. For example, if the physical absorption is simulated, then the equation (3.1) must be used to determine the optimal modification, where $k(x,x')$ is replaced with $k(x,x')q^{-1/2}(x)$. This conclusion can be obtained by introducing a new phase coordinate with two values: absorption and survival and replacing the first integral with the corresponding sum in (3.9).

Note also that (3.12) defines the optimal simulation of the total transition in the case of alternating signs, for example, if $h[2\varphi - h] \ge 0$, since then

$$\|G(g_1)]^{1/2} - [G(g_2)]^{1/2}\| \le q_0 \|g_1 - g_2\|,$$

i.e., there exists a unique solution of (3.12), Lemma 3.1 is true and (3.8) is satisfied.

In Sect. 3.6 we justify the utilization of (3.12) in a more general case where non-Markov chains $\{x_n\}$ are used while the optimal chain is Markovian.

3.3 A Model Problem

The equation

$$\varphi(z) = \begin{cases} q \int\limits_z^\infty \exp\{-(z'-z)\}\varphi(z')dz', & 0 \le z \le H, \\ h(z) \equiv 1, & z > H \end{cases} \tag{3.13}$$

can be solved by direct simulation of the following Markov chain.

A particle is moving on the axis making random jumps the length of which has the following distribution density: $\exp(-x)$, $(x > 0)$. At the end point (at z') of the jump the particle is absorbing with probability $\mathbf{P}(z') = 1 - q(z')$, where $q(z') = q$ if $z' < H$ and $q(z') = 0$ if $z' > H$. The Markov chain consists of the particle positions at the end of the jumps, i.e., before the choice of absorption or survival. The quantity $\varphi(z)$ for $z < H$ is interpreted as the probability that the particle escapes through the boundary $z = H$, provided the chain started at the point z. It is easy to verify that

$$\varphi(z) = q\exp\{-(1-q)(H-z)\}, \quad 0 < z < H.$$

For the chosen $\mathbf{P}(z')$, the optimal distribution density of the free length is obtained from (3.13) by replacing q with $q^{1/2}$ (Sect. 3.2):

$$g^{1/2}(z) = \begin{cases} q^{1/2}\exp\{-(1-q^{1/2})(H-z)\}, & 0 \le z \le H, \\ 1, & z > H. \end{cases}$$

Substituting this function in (3.10) gives the following distribution density of the free length of a jump starting at a point $z < H$:

$$f_l(x) = \begin{cases} q^{1/2}\exp\{-q^{1/2}x\}, & z + x \le H, \\ \exp\{-q^{1/2}H - (x-H)\}, & z + x > H. \end{cases}$$

In this case

$$\begin{aligned} \mathbf{M}\xi_z^{*2} &= g^2(z) = q\exp\{-2(1-q^{1/2})(H-z)\} \\ &\le q\exp\{-(1-q)(H-z)\} = \varphi(z), \quad z < H. \end{aligned}$$

Thus the optimal variant of simulation of the free length is indeed uniformly better than the direct simulation for which $\mathbf{M}\xi_z^2 = \varphi(z)$. Note that the expression for $\mathbf{M}\xi_z^2$ can be obtained by direct evaluations since the set of states of the Markov chain is in fact a homogeneous Poisson process on the interval $(0, H)$.

For comparison, let us consider the simulation of the free length according to (3.10) where $g^{1/2}$ is replaced with φ, i.e., the importance sampling technique (Sect. 2.1). The absorption is simulated as usual with probability $1 - q(z')$. The function $\mathbf{M}\xi_z^2$ is defined by

$$\mathbf{M}\xi_z^2 = \int\limits_z^H \exp\{-(2-q)(z'-z)\}\mathbf{M}\xi_{z'}^2\, dz' + q\exp\{-(2-q)(H-z)\},$$

where $z \leq H$. It is not difficult to verify that

$$\varphi(z) = q \exp\{-(1-q)(H-z)\}$$

is the unique solution of this equation. Thus the variance of the simulation according to the importance sampling technique in this case coincides with that of the direct simulation method.

3.4 Asymptotic Optimization of the Radiative Transfer

Let us consider the problem of calculating the probability of transmission of photons through an optically thick layer of medium $0 \leq z \leq H$. We assume that outside of the layer there is an absolute absorber and the free length σ^{-1} is equal to unity uniformly in the space. Escape through the plate $z = H$ is equivalent to a collision at $z > H$, for which $h = \varphi = \mathrm{M}\xi^2 \equiv 1$, otherwise $h = 0$. The scattering at a collision point is described by a symmetric density $w(\mu, \mu')$, where μ is the cosine of the angle between the direction of the run and the axis z; $q < 1$ is the survival probability after a collision inside the layer.

Let us first consider the optimization of simulation density of the free length l, the physical distribution of which is e^{-l}. Then in (3.9) $k_1(x, y) = qw(\mu, \mu')\delta(z_y - z)$ and

$$k_2(y, x) = |\mu|^{-1} \exp\{-(z' - z)/\mu\}\delta(\mu - \mu'), \quad (z' - z)/\mu' > 0.$$

The scattering and absorption are simulated according the physical laws, i.e., $p_1 \equiv k_1$. The equation (3.9) takes the form

$$g(z, \mu) = q \int_{-1}^{1} w(\mu, \mu') \left[\int_A^B |\mu'|^{-1} \exp\left\{ -\frac{z' - z}{\mu'} \right\} g^{1/2}(z', \mu') dz' \right]^2 d\mu,$$

where $g \equiv 1$ if $z > H$ and $(A, B) = (z, \infty)$ if $\mu' > 0$ and $(A, B) = (0, z)$ if $\mu' < 0$.

We approximate this equation by the equation for the infinite space. This approximation is based on the fact that the asymptotic solution to the linear transfer equation coincides [to within a constant factor which is not important in (3.10)] with the solution for the hole space. We seek the solution in the form $g^{1/2}(z, \mu) = a(\mu) \exp(cz)$. Simple arguments show that

$$a^2(\mu) = q \int_{-1}^{1} w(\mu, \mu') a^2(\mu')(1 - c\mu')^{-2} \, d\mu'. \tag{3.14}$$

As a solution to this equation, we use the transport approximation

$$w(\mu, \mu') = \mu_0 \delta(\mu - \mu') + (1 - \mu_0)/2,$$

where μ_0 is the mean scattering angle. The equation (3.14) is homogeneous; thus we can assume that $\int_{-1}^{1} a^2(\mu)\, d\mu = 1$. By integrating (3.14) we obtain, using the symmetry of $w(\mu, \mu')$, that

$$\int_{-1}^{1} a^2(\mu')(1 - c\mu')^{-1}\, d\mu' = q^{-1}.$$

By substituting into this equation the transport approximation to $a^2(\mu)$, we obtain the following equation for c:

$$\ln \frac{[c - (q\mu_0)^{1/2}]^2 - 1}{[c + (q\mu_0)^{1/2}]^2 - 1} = c\frac{4}{1 - \mu_0}[\mu_0/q]^{1/2}. \tag{3.15}$$

In the case of light scattering in sea water $q \simeq 0.7, \mu_0 \simeq 0.9$ [3.1, p. 182]; the last equation here gives $c \approx 0.206$ but the asymptotics φ is already close to its exponential behavior with $c_\varphi \approx 0.370$. In the case of isotropic scattering we have from (3.14) $c = (1 - q)^{1/2}$. When $q = 0.7$, this formula gives $c \approx 0.548$, while $c_\varphi \approx 0.829$.

In both cases the asymptotically optimal importance function $g^{1/2}$ is close to $\varphi^{1/2}$, but the first function change is slightly weaker than the second one. This is understandable since $\mathbf{M}\xi_x^2 = \varphi(x)$ for the direct simulation, while the optimal modification makes $\mathbf{M}\xi_x^2$ closer to $\varphi^2(x)$. Notice also that from (3.14) it follows that $c = 1 - q^{1/2}$ if $\mu = 1$. This corresponds to the exact solution to (3.9) obtained in Sect. 3.3 for a problem which is in fact related to the transmission of photons with δ-scattering through a layer of medium (i.e., when $\mu_0 = 1$).

To optimize the free length when simulating the trajectories without absorption, it is necessary to replace in (3.14) q with q^2 (Sect. 3.2). For $q = 0.7$ we get $\mu_0 = 0.9$, and from (3.15) we have $c \approx 0.336$.

It is known that the use of the function $g^{1/2}(z) = \exp(cz)$ in (3.10) leads to a modification of the distribution of the free length which consists of the transformation $\sigma' = \sigma - c\mu$, i.e., to the exponential transformation technique ([3.2] and Sect. 1.6). Thus the results obtained in this section can be interpreted as optimization of the exponential transformation.

This method also gives the asymptotics of the variance reduction compared to the direct simulation. It is clear that $\mathbf{M}\xi_z^2$ at $z = H$ is close to the unity in all variants. Therefore the variance $D\xi$ of the estimate for calculating the probability of light transmission through the layer $0 \le z \le H$ can be approximately taken as $cg(H)$. The factor c depends on the boundary effects and can be approximated by a constant in all the algorithms described below.

Algorithm 1 (Direct Simulation Where $\mathbf{M}\xi_z^2 = \varphi(z)$). We determined the asymptotics for $\varphi(z)$ on the basis of the transport approximation to $w(\mu, \mu')$, which leads to the isotropic scattering with $q' = q(1 - \mu_0)(1 - q\mu_0)^{-1}$ and $\sigma' = \sigma[q(1 - \mu_0) + 1 - q]$.

Algorithm 2 (Direct Simulation Without Absorption). $\mathbf{M}\xi_z^2$ satisfies (3.1) where q is replaced with q^2.

Algorithm 3 (Optimal Simulation of the Free Length With the Physical Absorption)

Algorithm 4 (Optimal Simulation of the Free Length Without Absorption)

Algorithm 5 (The Free Length Simulation According to the Density $q\exp(-qx)$ $(x > 0)$ Without Absorption). In this case the trajectory at the escape point has the weight

$$Q(L_z) = \exp\{-(1-q)L_z\},$$

where L_z is the total length of the trajectory started at the point z. It is obvious that the quantity

$$\mathbf{M}\xi_z^2 = \mathbf{M}Q^2(L_z) = \mathbf{M}\exp\{-2(1-q)L_z\}$$

satisfies (3.14) where $1-q$ is replaced with $2(1-q)$, i.e., when $q' = q/(2-q)$, $\sigma' = 2-q$.

We will now compare the costs of these algorithms $t\,D\xi$, where t is the effective average number of collisions per one trajectory. If we neglect the boundary effects, then for the simulation with absorption $t = 1/(1-q)$. It is clear that in the case of simulation without absorption it is possible to simulate $\ln\Phi \ln q$ collisions where Φ is the probability that a particle goes through the layer. Therefore as a relative cost of the algorithm without absorption we take

$$T = \frac{D\xi}{D\xi_0}\,\frac{\ln\,\Phi/\ln\,q+t}{t} = \frac{D\xi}{D\xi_0}\left[(1-q)\frac{\ln\Phi}{\ln q}+1\right],$$

where $D\xi_0$ is the variance of the direct simulation method. The asymptotic properties of the algorithms presented are illustrated by estimations obtained for $\mu_0 = 0.9, H = 20,(\ q = 0.7)$ (Table 3.1) and for $\mu_o = 0$, $H = 10$ (Table 3.2).

The results of Table 3.1 show that the algorithm 1 is almost optimal when the scattering has a large anisotropy. This is because the algorithm presents in fact the exact simulation according to the importance sampling without absorption and gives $D\xi = 0$ when $\mu_0 = 0$. It is interesting to note that, in the case of isotropic scattering, algorithm 2, which introduces a weight q which takes the absorption into account, is less effective than the direct simulation technique. The Monte Carlo calculations agree with the theoretical estimations shown in Tables 3.1, 2.

To optimize the distribution of μ' after the scattering it is sufficient to invert the order of choice of μ' and l. This means that x, the state after the scattering, is taken. Using arguments similar to that presented at the beginning of this section, we get the following equation for the function $a(\mu)$:

$$(1-2c\mu)^{1/2}\,a(\mu) = q^{1/2}\int_{-1}^{1} w(\mu,\mu')a(\mu')\,d\mu'. \tag{3.16}$$

Assuming that

$$\int_{-1}^{1} a(\mu)\, d\mu = 1,$$

we obtain in the case of a transport approximation

$$a(\mu) = \frac{q^{1/2}(1-\mu_0)}{2} \frac{1}{(1-2c\mu)^{1/2} - \mu_0 q^{1/2}},$$

where

$$c = \frac{(1-\mu_0)q^{1/2}}{2}\left[(1+2c)^{1/2} - (1-2c)^{1/2} + \mu_0 q \ln \frac{(1+2c)^{1/2} - \mu_0 q^{1/2}}{(1-2c)^{1/2} - \mu_0 q^{1/2}}\right].$$

In the case of isotropic scattering (i.e., when $\mu_0 = 0$) this equation takes the form

$$(1+2c)^{1/2} - (1-2c)^{1/2} = 2cq^{-1/2}.$$

This equation has a solution only if $q > 0.5$, since $(1+2c)^{1/2} - (1-2c)^{1/2} \le 2 \times 2^{1/2} c$ when $c < 0.5$.

For $q > 0.5$ we obtain

$$a(\mu) = (1-2c\mu)^{-1/2}, \quad c = [q(1-q)]^{1/2}$$

while the function $a_\varphi(\mu) = (1 - c_\varphi \mu)^{-1}$ corresponds to the asymptotics φ. For example, for $q = 0.7$ we have $2c = 0.458$, $c_\varphi = 0.370$.

Table 3.1

Quantity	Algorithm				
	1	2	3	4	5
$\ln D\xi/H$	0.37	0.559	0.412	0.672	0.67
$D\xi_0/D\xi$	1.0	44.0	2.3	420.0	403.0
T^{-1}	1.0	6.1	2.3	58.0	56.0

Table 3.2

Quantity	Algorithm				
	1	2	3	4	5
$-\ln D\xi/H$	0.829	0.961	1.096	1.428	1.223
$D\xi_0/D\xi$	1.0	3.7	14.4	399.0	51.0
T^{-1}	1.0	0.5	14.4	50.0	6.4

3.5 Asymptotic Optimization in a Special Class of Densities

In this section we construct and investigate nonlinear equations determining the optimal characteristics of the Markov chain in the case when the density for the angular distribution is taken from a special class which is convenient to simulate while the optimization is carried out with respect to the spatial variables. For details see [3.3].

We consider here, as in Sect. 3.4, the problem of calculating the radiation transfer through an optically thick layer of a medium. The survival probability after a collision q, the mean free length $l_0 = \sigma^{-1}$ and the angular phase function $w(\mu, \mu')$, which is fully defined by the scattering indicatrix $w_s(\mu_s)$, are given. Here μ_s is the cosine of the scattering angle (Sect. 1.3). The probability to be found is $I = \varphi(0,1)$, where φ is the solution of the integral transfer equation of the type (3.1) for a plane-parallel layer with

$$h(z,\mu) = \begin{cases} \exp\{-\sigma(H-z)/\mu\}, & \mu > 0, \\ 0, & \mu \leq 0. \end{cases}$$

Simulation of a trajectory consists of the successive simulation of the following random variables: the cosine μ' of the angle between the run direction and the axis z, the free length z' and a coordinate which is either absorption or survival.

Suppose that for a given survival probability q_p after a collision the following condition holds:

$$q^2/q_p < 1. \tag{3.17}$$

Then it follows from the results of Sect. 3.2 that the optimal densities of the random variables μ',z' are determined asymptotically from the equation

$$\left(1 + \frac{l_0}{k}\mu\right) a(\mu) = \frac{q}{\sqrt{q_p}} \int_{-1}^{1} w(\mu,\mu') a(\mu') d\mu', \quad k > l_0 \quad . \tag{3.18}$$

The free length is then simulated, as in the case of an exponential transformation technique with the parameter $1/k$ without an escape, and the scattering is simulated according to the density

$$\frac{w(\mu,\mu')a(\mu')}{\int_{-1}^{1} w(\mu,\mu')a(\mu')d\mu'}. \tag{3.19}$$

Remark. We shall assume that $0 < M_1 \leq a(\mu) \leq M_2 < \infty$. This is satisfied, in particular, if the function $w(\mu, \mu')$ satisfies an analogous condition.

In the case of the anisotropic scattering it is difficult to simulate according to the density (3.19). Therefore it is convenient to replace the density (3.19) with the density of the type

$$pw(\mu,\mu') + (1-p)a(\mu'), \quad p = p(z,\mu), \quad 0 \le p \le 1. \tag{3.20}$$

In this section we solve the following important problem: for a fixed value of q_p satifying (3.17) find a density $p(z,\mu)$ in (3.20) and the density $r = r(z \to z'|\mu')$ of z' in such a way that the quantity $\mathbf{M}\xi_x^2$ is uniformly minimal. The Markov chain which corresponds to this modification is called optimal and is denoted by (r^*, p^*).

By substituting the values of the transition densities, we find that (3.3) takes the form (for arbitrary probability p and density r)

$$\begin{aligned}\psi(z,\mu) &= h(z,\mu)[2\varphi(z,\mu) - h(z,\mu)] \\ &\quad + \frac{q^2}{q_p}\int_{-1}^{1} \frac{w^2(\mu,\mu')}{p(z,\mu)w(\mu,\mu') + (1-p(z,\mu))a(\mu')} \\ &\quad \times \frac{1}{\mu'^2 l_0}\int_A^B \frac{\exp\{-2(z'-z)/(\mu' l_0)\}}{r(z-z'|\mu')}\Psi(z',\mu')dz'\,d\mu',\end{aligned} \tag{3.21}$$

or

$$\Psi = h(2\varphi - h) + u(r,p),$$

where $(A,B) = (z,H)$ if $\mu > 0$ and $(A,B) = (0,z)$ if $\mu < 0$.

We shall also assume that $\tilde{\Psi} \le c_1\varphi^*$, where $\tilde{\Psi}$ is the Neumann series for the equation

$$\tilde{\Psi} = h(2\varphi - h) + u(k_l, 1)\tilde{\Psi},$$

$$k_l(z \to z'|\mu') = \frac{1}{|\mu'| l_0}\exp\left\{-\frac{z'-z}{\mu' l_0}\right\}.$$

Now we fix $\Psi(z',\mu')$ in the inner integral of (3.21) and vary the density of the transition $(z,\mu') \to (z',\mu')$. By the importance sampling principle this integral is minimal and equals

$$\left[\int_B^A k_l(z,z'|\mu')\Psi^{1/2}(z',\mu')dz'\right]^2$$

if

$$r(z \to z'|\mu) = k_l(z,z'|\mu')\Psi^{1/2}(z',\mu')\left[\int_A^B k_l(z,z'|\mu')\Psi^{1/2}(z',\mu')dz'\right]^{-1}.$$

Let us fix the point (z,μ) and let

$$F(p) = \int_{-1}^{1} \frac{w(\mu, \mu')}{p(z, \mu)w(\mu, \mu') + (1 - p(z, \mu))a(\mu')} \times \left[\int_{A}^{B} k_l(z, z'|\mu')\Psi^{1/2}(z', \mu')dz' \right]^2 d\mu',$$

or

$$F(p) = V(p)\Psi.$$

Since the function $a > 0$ is bounded, we conclude that for arbitrary $\Psi \in L_\infty(X)$ the function $F(p)$ is k-times differentiable on $[0, 1]$, where $k = 1, 2 \ldots$. In addition, $F'' > 0$ if $w \neq a$. Therefore, there exists a unique function $p_0(z, \mu)$, which minimizes $F_0 = F$ [3.4]. Thus

$$\Psi_1(z, \mu) = h(z, \mu)[2\varphi(z, \mu) - h(z, \mu)] + \frac{q^2}{q_p} \int_{-1}^{1} \frac{w(\mu, \mu')}{p(z, \mu)w(\mu, \mu') + (1 - p(z, \mu))a(\mu')} \times \left[\int_{A}^{B} k_l(z, z'|\mu')\Psi_0^{1/2}(z', \mu')dz' \right]^2 d\mu' \leq \Psi_0(z, \mu) = \Psi(z, \mu), \tag{3.22}$$

or

$$\Psi_1 = h(2\varphi - h) + G(p_0)\Psi_0 \leq \Psi_0.$$

By replacing Ψ_1 with Ψ_0 and minimizing the function $F_1(p) = V(p)\Psi_1$, we get

$$\Psi_2 = h(2\varphi - h) + G(p_1)\Psi_1 \leq \Psi_1, \quad p_1 = \min_p F_1(p).$$

Thus we obtained a monotonically decreasing sequence of positive functions

$$\Psi_{n+1} = h(2\varphi - h) + G(p_n)\Psi_n, \quad p_n = \min_p F_n(p), \quad n = 0, \ldots \quad . \tag{3.23}$$

The pointwise convergence

$$\Psi_n(x) \to \Psi^*(x),\ F_n(p)|_x \to F^*(p)|_x, \quad p_n(x) \to p^*(x) \quad \forall x \in X$$

is obvious. Here $F^*(p) = V(p)\Psi^*, \quad p^* = \min_p F^*(p)$. Taking a limit in (3.23) as $n \to \infty$, we get

$$\Psi^*(z, \mu) = h(z, \mu)[2\varphi(z, \mu) - h(z, \mu)] + \frac{q^2}{q_p} \int_{-1}^{1} \frac{w^2(\mu, \mu')}{p^*(z, \mu)w(\mu, \mu') + (1 - p^*(z, \mu))a(\mu')} \times \left[\int_{A}^{B} k_l(z, z'|\mu')\Psi^{*1/2}(z', \mu')dz' \right]^2 d\mu', \tag{3.24}$$

$$p^* = \min_p \int_{-1}^{1} \frac{w^2(\mu,\mu')}{p(z,\mu)w(\mu,\mu') + (1-p(z,\mu))a(\mu')} \times \left[\int_A^B k_l(z,z'|\mu')\Psi^{*1/2}(z',\mu')dz' \right]^2 d\mu,$$

or

$$\Psi^* = h(2\varphi - h) + G(p^*)\Psi^*, \quad p^* = \min_p V(p)\Psi^*.$$

The equation derived is nonlinear. The statements which follow show how to define the optimal Markov chain.

Lemma 3.4. If $\Psi^* \in L_\infty(X)$ is the solution to (3.24), then for the chain with the characteristics

$$r \equiv r^*(z \to z'|\mu) = k_l(z,z'|\mu')\Psi^{*1/2}(z',\mu') \left[\int_A^B k_l(z,z'|\mu')\Psi^{*1/2}(z',\mu')dz' \right]^{-1},$$

$$p(z,\mu) = p^*(z,\mu)$$

the following relation holds:

$$\mathrm{M}\xi_x^2 = \Psi^*(x).$$

Lemma 3.5. If Ψ^* is the solution to (3.24), then the following inequality holds:

$$\varphi^2 \le \Psi^* \le \varphi. \tag{3.25}$$

Lemma 3.6. The solution to (3.24) is unique in $L_\infty(X)$.

These Lemmata could be proved like the Lemmata of Sect. 3.2.

Theorem 3.2. There exists a unique solution to (3.24) in $L_\infty(X)$. The corresponding Markov chain (r^*, p^*) is optimal.

Proof. The existence follows from the fact that $\lim \Psi_n$ at $\Psi_0 = \Psi$ satisfies (3.24). The uniquness is given by Lemma 3.6. The chain with $r \equiv r^*$, $p \equiv p^*$ is optimal, since the convergence

$$\Psi_n(x) \downarrow \Psi^*(x) \quad \forall \Psi_0 \in L_\infty(X)$$

is monotonic. □

It is known [3.5] that the asymptotic solution of the transfer equation coincides, to within a constant factor, with the solution g in the whole space. Therefore, we approximate (3.24) by the corresponding equation in the whole space. It is natural to assume then that the optimal probability p^* weakly depends on the coordinate z of the collision point and depends only on μ.

Simple arguments show that after substituting

$$\Psi^{*1/2}(z,\mu) = b(\mu)\exp(cz) \quad ,$$

we obtain the following equation:

$$b^2 = \frac{q^2}{q_p} F(b^2, c, p^*), \tag{3.26}$$

where

$$F(b^2,c,p)|_\mu = \int_{-1}^{1} \frac{w^2(\mu,\mu')}{p(\mu)w(\mu,\mu') + (1-p(\mu))a(\mu')} \frac{b^2(\mu')}{(1-cl_0\mu')^2}\, d\mu',$$

$$p^* = \min_p F(b^2, c, p).$$

If we suppose that $p^*(\mu) \neq 0,1$, then the values of $p^*(\mu)$ from the interval $(0,1)$ satisfy the relation

$$\int_{-1}^{1} \frac{w^2(\mu,\mu')}{p(\mu)w(\mu,\mu') + (1-p(\mu))a(\mu')} \frac{b^2(\mu')}{(1-cl_0\mu')^2}(w(\mu,\mu') - a(\mu'))d\mu' = 0 \quad .$$

Let us now study the approximate solution (3.26) in the case when the scattering indicatrix has a sharp peak. We show that if $\mu_0 \simeq 1$, then the approximate asymptotically optimal chain (r_0, p_0) is defined as follows: the density $r_0(z \to z'|\mu)$ corresponds to the exponential transformation (Sect. 1.3) with the parameter $c = c_0 = 1/k$, and $p(\mu)$ satisfies the equation

$$\int_{-1}^{1} \frac{w^2(\mu,\mu')a^2(\mu,\mu')}{[p(\mu)w(\mu,\mu') + (1-p_0(\mu))a(\mu')]^2}(w(\mu,\mu') - a(\mu'))d\mu' = 0. \tag{3.27}$$

Indeed, the change of variables

$$b(\mu) = (1 - cl_0\mu)\theta(\mu)$$

in (3.26) yields

$$\theta^2(\mu) = \frac{q^2}{q_p} \frac{1}{(1-cl_0\mu)^2} \Phi(\theta^2, p^*),$$

where

$$\Phi(\theta^2, p)|_\mu = \int_{-1}^{1} \frac{w^2(\mu,\mu')\theta^2(\mu')}{p(\mu)w(\mu,\mu') + (1-p(\mu))a(\mu')}\, d\mu',$$

$$p^* = \min_p \Phi(\theta^2, p).$$

Using the transport approximation

$$w(\mu,\mu') = \mu_0\delta(\mu-\mu') + (1-\mu_0)/2$$

we obtain

$$\Phi(a^2, p_0) = (\sqrt{\mu_0}a + (1-\mu_0)/2)^2.$$

Here the terms of order $(1-\mu_0)^2$ are neglected.

From this we find that for μ_0 close to 1 the following approximate equality holds:

$$\Phi(a^2, p_0) \approx \left[\int_{-1}^{1} w(\mu,\mu')a(\mu')\, d\mu'\right]^2 . \tag{3.28}$$

Thus, according to the importance sampling principle, p_0 is an approximate minimum of the function $\Phi(a^2, p)$.

In addition, from (3.18, 28) it follows that (3.26) is approximately valid at $\theta \equiv a$, $p^* \equiv p_0$.

We now calculate the probability I_H of a transmission of particles through a layer $0 \le z \le H$ for $H = 20l_0$, $l_0 = 1$. While the scattering indicatrix is taken as

$$w(\mu_s) = \frac{1}{2}\frac{1-\mu_0^2}{(1+\mu_0^2 - 2\mu_0\mu_s)^{3/2}},$$

the mean cosine of the scattering angle $\mu_0 = 0.99$, the absorption probability after a collision $1-q = 0.1$. The source density is $f(z,\mu) = \delta(\mu-1)\delta(z)$.

The survival probability q_p was chosen as follows:

$q_p = 1$ if $n < m$, $q_p = 0.901$ if $n \ge m$, where $m = |\ln g(0,1)/\ln q| = 14$.

Using the transport approximation for determining $a(\mu)$ and k, we obtain from (3.18)

$$a(\mu) = \frac{\tilde{q}}{2}\frac{k}{k-\tilde{l}\mu} .$$

Here

$$\tilde{q} = \frac{\hat{q}(1-\mu_0)}{1-\hat{q}\mu_0}, \quad \tilde{l} = \frac{l}{1-\hat{q}\mu_0}, \quad \hat{q} = \frac{q}{\sqrt{q_p}} ,$$

and k is the solution to the equation

$$\frac{2\tilde{l}}{\tilde{q}k} = \ln \frac{k+\tilde{l}}{k-\tilde{l}} \quad .$$

We obtain

$$\hat{q} = 0.9, \quad \tilde{q} = 0.4737, \quad \tilde{l} = 5.2632, \quad k = 5$$

for $n < m$ and

$$\hat{q} = 0.95, \quad \tilde{q} = 0.6552, \quad \tilde{l} = 6.8966, \quad k = 7.9$$

for $n \geq m$.

Remark. From the definition of $w(\mu, \mu')$ it follows that

$$w(1, \mu') = w_s(\mu'), \quad w(-1, \mu') = w_s(-\mu').$$

Therefore the scattering according to the density (3.19) in cones $K_{\pm\varepsilon} = \{\pm\mu > 1 - \varepsilon\}$ can be approximately carried out according to the densities

$$f^{\pm}(\mu') = \frac{w_s(\pm\mu')a(\mu')}{\int_{-1}^{1} w_s(\pm\mu')a(\mu')d\mu'}.$$

The calculations shows that the optimal value of ε is close to 0.12. A comparison of the algorithms is presented in the Tables 3.3, 4.

Table 3.3

Specification	Algorithm		
	A	B	C
The free length is simulated without escape as in the exponential transformation.	$\frac{1}{k}$	$\frac{1}{L}$	$\frac{1}{L}$
Scattering is simulated from the density.			
$-0.88 \leq \mu \leq 0.88$	$p_0 w + (1 - p_0)a$	$p_1 w + (1 - p_1)a_2$	$p_2 w + (1 - p_2)a_0$
$0.88 \leq \mu$	f^+	f_0^+	$p_2 w + (1 - p_2)a_0$
$\mu \leq -0.88$	f^-	f_0^-	$p_2 w + (1 - p_2)a_0$

Table 3.4

Algorithm	V_H	$\sigma, \%$	t, c	σ^2, t
A	$1.9439 \cdot 10^{-2}$	2.95	$5.71 \cdot 10^2$	0.4960
B	$1.9424 \cdot 10^{-2}$	3.19	$5.62 \cdot 10^2$	0.5710
C	$1.9398 \cdot 10^{-2}$	3.96	$5.47 \cdot 10^2$	0.8568

Here $L = 5.4$ is the diffusion length [3.5, 6], $p_0(\mu)$ is the solution to (3.27), the functions $a_0(\mu)$, $p_1(\mu)$ are defined to be similar to $a(\mu)$, $p_0(\mu)$ where $\hat{q}$ is replaced with q, and

$$f_0^{\pm}(\mu) = w_s(\pm\mu)a_0(\mu)\left[\int_{-1}^{1} w_s(\pm\mu)a_0(\mu)\,d\mu\right]^{-1}.$$

The density $p_2(\mu)$ is defined by (2.24), which corresponds to the simple approximation of the function $w(\mu,\mu')a_0(\mu')$ described in Sect. 2.4. Notice that $k = L$ for $n < m$ and $k > L$ for $n \geq m$.

In each variant 10^4 trajectories were simulated. In the tables the following notations are used: Y_H is the estimate of the functional I_H, σ is the estimate of the relative statistical error, t is the computer time (in seconds). The calculations show that the relative costs of algorithms A, B and C relate as 58:67:100, respectively.

3.6 Minimization of the Variance of the Collision Estimates

As in Sect. 3.1, we will now consider the equation

$$\varphi(x) = \int_X k(x,x')\varphi(x')dx' + h(x), \quad x,x' \in X, \tag{3.29}$$

or in the operator form

$$\varphi = K\varphi + h,$$

where X is the n-dimentional Euclidian space, $h \in L_\infty(X), K \in L_\infty$. Let $x_0,\ldots,x_n,\ldots$ be a trajectory of a random process in X, which is defined by the transition densities $p_n(x_0,\ldots,x_{n-1},x) = [1-p(x_{n-1})] \times r_n(x_0,\ldots,x_{n-1},x)$, where $r_n(x_0,\ldots,x_{n-1},x)$ is the probability density at the n-th step, and $p(x_{n-1})$ is the terminating probability when going from the point x_{n-1}. We now define the random weights

$$Q_n = Q_{n-1}\delta_{n-1}\frac{k(x_{n-1},x_n)}{p_n(x_0,\ldots,x_{n-1},x_n)}, \quad Q_0 = 1, \quad n \geq 1,$$

where $\delta_{n-1} = 0$ if at the n-th step the trajectory breaks, and $\delta_{n-1} = 1$ otherwise. We assume in what follows that
(a) $p_n(x_0,\ldots,x_{n-1},x_n) \neq 0$ for $k(x_{n-1},x_n)\varphi(x_n) \neq 0$
(b) the spectral radius of the integral operator K_1 with the kernel $|k(x,x')|$ is less than unity: $\rho(K_1) < 1$.

It is known (Sect. 1.6) that under these conditions the collision estimate

$$\xi_{x_0} = \sum_{k=0}^{\infty} Q_k h(x_k) \tag{3.30}$$

is an unbiased esimate for the solution to (3.29) at the point x_0, i.e., $\mathbf{M}\xi_{x_0} = \varphi(x_0)$. Here ξ_{x_0} is considered to be a random variable defined on a space of trajectories starting at x_0, and the expectation is taken over the probabilistic measure generated by the transition probabilities. For simplicity we will omit the index x_0. Thus we simply write $\mathbf{M}\xi = \varphi$ or $\mathbf{M}\xi^2 \geq g$ if these relations hold almost everywhere in X (with respect to the Lebesgue measure).

Now we fix the density $p(x)$ for all $x \in X$ and try to find the densities $r_n(x_0, \dots, x_{n-1}, x)$ so that the variance of ξ_{x_0} would be minimal uniformly over x_0. In Sect. 3.2 we solved the problem where the kernel $k(x, x')$ and the function $h(x)$ are nonnegative. The results can be generalized to the case when k and h have alternating signs if the following condition holds: $h(2\varphi - h) \geq 0$. This condition is equivalent to the assumption that $|\varphi| \geq |K\varphi|$, which is limiting. For example it is not satisfied in the diffraction problems. In addition, in Sect. 3.2 we assumed that $\|K\| < 1$ and $p(x) \leq p < 1 - \|K\|^2$. We now consider another approach which permits us to solve the problem in the general case.

Let

$$\eta^{(0)} = \varphi(x_0), \quad \eta^{(n)} = \sum_{k=0}^{n-1} Q_k h(x_k) + Q_n \varphi(x_n), \quad n \geq 1.$$

For arbitrary n the random quantity $\eta^{(n)}$ is in fact a conditional expectation of the random variable ξ and all $\eta^{(n+k)}$ $(k \geq 0)$ for fixed $x_1, \dots, x_n$, i.e., the sequence $\eta^{(n)}$ is a martingal which is closed from the right by ξ. Since $\eta^{(n)}$ is a martingal, $(\eta^{(n)})^2$ is a submartingal. Thus the sequence $\mathbf{M}[\eta^{(n)}]^2$ is increasing and has a limit (possibly an infinite one) which is less than $\mathbf{M}\xi^2$ [3.7]. On the other hand, we can conclude that the sequence $\eta^{(n)}$ converges to ξ with probability one, since

$$\mathbf{M}|\xi - \eta^{(n)}| = \mathbf{M}|Q_n\varphi(x_n) + \sum_{k=n}^{\infty} Q_k h(x_k)| \leq K_1^n|\varphi| + \sum_{k=n}^{\infty} K_1^k|h|$$

and

$$K_1^n|\varphi| + \sum_{k=n}^{\infty} K_1^k|h| \to 0, \quad n \to \infty$$

[3.7]. From this we get

$$\mathbf{M}\xi^2 = \lim \mathbf{M}\left\{\inf_{k\geq n}[\eta^{(k)}]^2\right\} \leq \lim \mathbf{M}[\eta^{(n)}]^2.$$

Thus the following relation holds:

$$\mathbf{M}\xi^2 = \lim \mathbf{M}[\eta^{(n)}]^2. \tag{3.31}$$

Notice that ξ can be represented in the form

$$\xi_{x_0} = h(x_0) + Q_1 \tilde{\xi}_{x_1}, \tag{3.32}$$

where the random variable $\tilde{\xi}$ is defined as in (3.30) by the sequence of the densities $r_2(x_0, x_1, x), \ldots, r_n(x_0, \ldots, x_{n-1}, x)$. So it may depend, in the general case, on the parameter x_0. We fix the points $x_0, \ldots, x_n$ and integrate (3.32) over the rest of the variables. This yields

$$\eta_{x_0}^{(n)} = h(x_0) + Q_1 \tilde{\eta}_{x_1}^{(n-1)}, \tag{3.33}$$

where $\tilde{\eta}_{x_1}^{(n-1)}$ is the conditional expectation of the random variable $\tilde{\xi}_{x_1}$ under the condition that $x_2, \ldots, x_n$ are fixed.

We now define the sequence of functions g_n as follows:

$$g_n(x) = \inf \mathbf{M}[\eta_x^{(n)}]^2, \quad n \geq 0,$$

where the inf is taken over all sequences of densities $r_n(x_0, \ldots, x_{n-1}, x)$ satisfying the condition (a). The sequence g_n increases (pointwise) and has a limit (finite or infinite), since

$$\mathbf{M}[\eta^{(n)}]^2 \geq \mathbf{M}[\eta^{(n-1)}]^2 \geq g_{n-1}.$$

Let $g = \lim g_n$. From (3.33) it follows that

$$\begin{aligned} \mathbf{M}[\eta_{x_0}^{(n)}]^2 &= \mathbf{M}[h(x_0) + Q_1 \tilde{\eta}_{x_1}^{(n-1)}]^2 \\ &= h(x_0)[2\varphi(x_0) - h(x_0)] + \frac{1}{1-p(x_0)} \int_X \frac{k^2(x_0, x_1)}{r_1(x_0, x_1)} \mathbf{M}[\tilde{\eta}_{x_1}^{(n-1)}]^2 \, dx_1. \end{aligned} \tag{3.34}$$

This shows, in particular, that if the densities $r_n(x_0, \ldots, x_{n-1}, x)$ are fixed, the increase of the absorption probability $p(x)$ always leads to the increase of the variance of ξ.

Notice that $\mathbf{M}[\tilde{\eta}^{(0)}]^2 = \varphi^2 = g_0$ for arbitrary $\tilde{\xi}$. Let $\tilde{\xi}$ be a random variable such that $\mathbf{M}[\tilde{\eta}^{(n-1)}]^2 = g_{n-1}$ for an integer n. Then from the importance sampling technique it follows (Sect. 1.5) that for the random variable $\xi_{x_0} = h(x_0) + Q_1 \tilde{\xi}_{x_1}$ with the density $r_1(x_0, x)$ proportional to $|k(x_0, x)| g_{n-1}^{1/2}(x)$ the following relation holds: $\mathbf{M}[\eta^{(n)}]^2 = g_n$, where

$$g_n = h(x)[2\varphi(x) - h(x)] + \frac{1}{1-p(x)} \left[\int_X |k(x, x')| g_{n-1}^{1/2}(x') \, dx' \right]^2. \tag{3.35}$$

Taking the limit as $n \to \infty$ yields

$$g(x) = h(x)[2\varphi(x) - h(x)] + \frac{1}{1-p(x)} \left[\int_X |k(x, x')| g^{1/2}(x') \, dx' \right]^2. \tag{3.36}$$

This equation was studied in Sect. 3.2. In particular, it was proved that under the above assumptions (3.36) has a unique solution in $L_\infty(X)$.

By virtue of (3.31)

$$\mathbf{M}\xi^2 = \lim \mathbf{M}[\eta^{(n)}]^2 \geq \lim g_n = g \tag{3.37}$$

for arbitrary ξ.

Theorem 3.3. If the function g is finite almost everywhere in X, then the relations

$$\begin{aligned} & r_n(x_0, \ldots, x_{n-1}, x) \\ & \quad = |k(x_{n-1}, x)| g^{1/2}(x) \left[\int_X |k(x_{n-1}, x)| g^{1/2}(x) dx \right]^{-1}, \quad n \geq 1 \end{aligned} \tag{3.38}$$

determine the unique estimate ξ whose variance is equal to $g - \varphi^2$.

Proof. Using (3.38) we rewrite (3.24) in the form

$$\begin{aligned} \mathbf{M}[\eta_{x_0}^{(n)}]^2 &= h(x_0)[2\varphi(x_0) - h(x_0)] \\ &+ \frac{1}{1 - p(x_0)} \left[\int_X |k(x_0, x_1)| g^{1/2}(x_1)\, dx_1 \int_X |k(x_0, x_1)| \frac{\mathbf{M}[\eta_{x_1}^{(n-1)}]^2}{g^{1/2}(x_1)}\, dx_1 \right]. \end{aligned}$$

A comparison of this relation to (3.36) shows that from $\mathbf{M}[\eta^{(n-1)}]^2 \leq g$ it follows that $\mathbf{M}[\eta^{(n)}]^2 \leq g$ for arbitrary $n \geq 0$. If $n = 0$, the inequality $\mathbf{M}[\eta^{(0)}]^2 \leq g$ is obvious. Consequently, $\lim \mathbf{M}[\eta^{(n)}]^2 \leq g$; thus from (3.37) we get $\mathbf{M}\xi^2 = g$.

We now prove the uniqueness. From (3.32) we get

$$\begin{aligned} \mathbf{M}\xi_{x_0}^2 &= h(x_0)[2\varphi(x_0) - h(x_0)] \\ &\quad + \frac{1}{1 - p(x_0)} \left[\int_X \frac{k^2(x_0, x_1)}{r_1(x_0, x_1)} \mathbf{M}\tilde{\xi}_{x_1}^2\, dx_1 \right]. \end{aligned}$$

Since $\mathbf{M}\tilde{\xi}^2 \geq g$, it follows from the importance sampling principle that $\mathbf{M}\xi^2 = g$ holds only if the density $r_1(x_0, x)$ is proportional to $|k(x_0, x)| g^{1/2}(x)$ and $\mathbf{M}\tilde{\xi}^2 = g$. Analogous arguments for $\tilde{\xi}$ show that the density $r_2(x_0, x_1, x)$ must also satisfy (3.38) etc. □

Corollary. If $p(x) = 0$ for almost all $x \in X$, and the product $k(x, x')\varphi(x')$ as a function of x' is of constant signs, then $g = \varphi^2$; hence (3.38) define an estimate which has zero variance (i.e., it is identically equal to φ).

Indeed,

$$h(x)[2\varphi(x) - h(x)] = \varphi^2(x) - \left[\int_X k(x,x')\varphi(x')\,dx'\right]^2 .$$

Thus from (3.35) regarded for $n = 1, 2, \ldots$, we find that $g_n = \varphi^2$ for all n.

Theorem 3.4. If vrai sup$\{p(x) : x \in X\} = p < 1 - [\rho(K_1)]^2$, then $g \in L_\infty(X)$.

Proof. Let s be a positive integer such that $p < 1 - (\|K_1^s\|^{1/s})^2$. We now consider an estimate ξ which is defined by a periodic (with period s) sequence of densities $r_n(x_0, \ldots, x_{n-1}, x)$ which are proportional to

$$|k(x_{n-1}, x)| \int_X \cdots \int_X |k(x_0, x_1) \ldots k(x_{n-2}, x_{n-1})|\, dx_1 \ldots dx_{n-1},$$

for $n \leq s$ and $r_{ms+n}(x_0, \ldots, x_{ms+n-1}, x) = r_n(x_{ms}, \ldots, x_{ms+n+1}, x)$ for $m \geq 1$. It is not difficult to verify that the norm [in $L_\infty(X)$] of the integral operator $\tilde{K}_p$ with the kernel

$$\int_X \cdots \int_X \frac{k^2(x_0, x_1)}{p_1(x_0, x_1)} \cdots \frac{k^2(x_{s-1}, x)}{p_s(x_0, \ldots, x_{s-1}, x)}\, dx_1 \ldots dx_{s-1}$$

is less than unity.

It can be shown, as in [3.2], that in this case $\mathbf{M}\xi^2$ is the Neumann series of the equation $\psi = \tilde{K}_p\psi + \tilde{h}$, where

$$\tilde{h}(x) = \sum_{k=0}^{s-1} \int_X \cdots \int_X \frac{k^2(x_0, x_1)}{p_1(x_0, x_1)} \cdots \frac{k^2(x_{k-1}, x_k)}{p_k(x_0, \ldots, x_{k-1}, x_k)} h(x_k)[2\varphi(x_k) - h(x_k)] dx_1 \ldots dx_k,$$

$\tilde{h} \in L_\infty(X)$. Thus $\mathbf{M}\xi^2 \in L_\infty(X)$ and the inequality $g \leq \mathbf{M}\xi^2$ completes the proof. □

Remark. Assume that for arbitrary x

$$h(x)[2\varphi(x) - h(x)] > 0,$$

$$|\varphi(x)| > \epsilon \geq 0, \qquad \int_X |k(x,x')|\, dx' = q < 1$$

and $\rho(K_1) = q$. We put $p(x) = 1 - q^2$ for all $x \in X$. Then from (3.35) we see that g_n increases infinitely. Thus the inequality of Theorem 3.4 cannot be weakened.

We now generalize some of the results obtained for the nonlinear equation

$$\varphi(x) = \int\limits_X \cdots \int\limits_X k(x, y_1, \ldots, y_m) \prod_{i=1}^{m} \varphi(y_i) dy_i + h(x), \tag{3.39}$$

or in the operator form

$$\varphi = \mathcal{K}\varphi^{(m)} + h.$$

Let $\mathcal{K}_1$ be the integral operator with the kernel $|k(x, y_1, \ldots, y_m)|$. We assume that the method of successive approximations $\tilde{\varphi}_0 = |h|, \quad \tilde{\varphi}_{n+1} = \mathcal{K}_1\tilde{\varphi}_n^{(m)} + h$ for the equation $\tilde{\varphi} = \mathcal{K}_1\tilde{\varphi}^{(m)} + |h|$ converges.

Let ξ be an analog of the collision estimate for (3.39) [3.2]. We assume that the absorption probability $p(x)$ is fixed and now try to find an optimal transition density $r_n(x, y_1, \ldots, y_m)$.

Let $\eta^{(n)}$ be the conditional expectation of ξ under the condition that the first $(n+1)$–generations of the tree are fixed. The equality $\mathbf{M}\xi^2 = \lim \mathbf{M}[\eta^{(n)}]^2$ is proved analogously to (3.31).

Let

$$\begin{aligned} g_0 &= \varphi^2, \\ g_n(x) &= h(x)[2\varphi(x) - h(x)] \\ &\quad + \frac{1}{1 - p(x)} \left[\int\limits_X \cdots \int\limits_X |k(x, y_1, \ldots, y_m)| \prod_{i=1}^{m} g_{n-1}^{1/2}(y_i) dy_i \right]^2, \quad n \geq 1, \end{aligned}$$

and $g = \lim g_n$.

Theorem 3.5. If the function g is finite almost everywhere in X, then the densities $r_n(x, y_1, \ldots, y_m)$ proportional to $|k(x, y_1, \ldots, y_m)| \prod_{i=1}^{m} g^{1/2}(y_i)$ define the unique estimate ξ which has a minimal variance equal to $g - \varphi^2$.

Proof. The proof is almost exactly the same as that of Theorem 3.3. □

Corollary. If $p^m(x) = 0$ for almost all $x \in X$ and if the product

$$k(x, y_1, \ldots, y_m) \prod_{i=1}^{m} \varphi(y_i),$$

as a function of m variables $y_1, \ldots, y_m$, has a constant sign almost everywhere in X^m, then $g = \varphi^2$ and hence there exists an estimate ξ with zero variance.

Note that the results of this section could be generalized to the absorption estimate for linear and nonlinear cases by replacing $h(2\varphi - h)$ with h^2/p everywhere. In conclusion we note that the nonlinear optimization gives a possibility of investigating the algorithms (to minimize the variance) in the cases when zero-variance estimates do not exist. As an example, we mention the collision estimate for the integral equation with alternating elements. In particular, the exact theory of exponential transformation and a correction of the parameters of the approximate asymptotic importance function for the one-velocity transfer equation were constructed.

4. Minimax Weighted Estimates

As was already mentioned in Chap. 1, one of the most impressive features of the Monte Carlo methods is the possibility of solving a set of problems using the same probabilistic distribution. To make the corresponding estimates unbiased, one introduces special weights. Therefore, these are called weighted estimates. A characteristic feature of this approach is the possibility of constructing dependent estimates which improves the calculation of functions [4.1].

The weighted estimates improve the algorithms, especially if computers with parallel processors are used. Thus it is important to solve the problem of optimal choice of the distribution to be simulated. In this chapter we solve the problem of uniform optimization of the weighted Monte Carlo estimates which reduces to the minimization of the maximum of the variances. We give an exact and a special approximate solution in the case when a set of integrals must be calculated. An analogous exact solution is also constructed for the standard estimates for the functionals of the solutions to integral equations. It is shown how to carry out an approximate uniform optimization of the weighted estimates when solving a set of integral equations and the radiative transfer equation. In particular, we improved the method of similar trajectories.

It should be noted that we use the optimization of weighted Monte Carlo estimates according to criterion based on the weighted sum of the variances (Sects. 1.5, 1).

4.1 Statement of the Problem. The Basic Lemma

The Monte Carlo calculations of integrals are based on the relation

$$I_k = \int_X g_k(x)\,dx = \mathrm{M}\zeta_k, \quad \zeta_k = \frac{g_k(\xi)}{p(\xi)}, \quad k = 1,\ldots,n,$$

where $p(x)$ is the distribution density of ξ, and

$$D\zeta_k(p) = \int_X \frac{g_k^2(x)\,dx}{p(x)} - I_k^2, \quad F_\zeta(p) = \max_k\{D\zeta_k(p)\}.$$

We consider the following problem: find a density $p = p^*(x)$ such that $\min_p F_\zeta(p) = F_\zeta(p^*)$ is attained. We formulate the known [4.2] relation between the minimax and the Bayes solutions by making the following statement.

Lemma 4.1. Let $\xi_k(P)$ be random variables in a probabilistic model P of a class of admissible models, and let P_Λ be a model such that the minimum

$$\min \sum_k \lambda_k D\xi_k(P) = G(\Lambda), \quad 0 \le \lambda_k < \infty, \Lambda = (\lambda_1, \ldots, \lambda_n) \tag{4.1}$$

is attained, where $G(\Lambda) < \infty$ and $D\xi_k(P_\Lambda)$ are differentiable with respect to λ_i, $(i, k = 1, \ldots, n)$ for all Λ. Then $\min F_\xi(P)$ is attained in P_{Λ_0}, where Λ_0 is the solution to the problem

$$\max_\Lambda \left\{ \sum_{k=1}^n \lambda_k D\xi_k(P) | \quad \sum_{k=1}^n \lambda_k = 1 \right\}. \tag{4.2}$$

Proof. Since P_Λ solves (4.1), we get

$$\sum_k \lambda_k \frac{\partial D\xi_k(P_\Lambda)}{\partial \lambda_i} \begin{cases} = 0, & 0 < \lambda_i < 1, \\ \ge 0, & \lambda_i = 0, \\ \le 0, & \lambda_i = 1. \end{cases} \tag{4.3}$$

We differentiate the function

$$G(\lambda_1, \ldots, \lambda_n) = \sum_k \lambda_k D\xi_k(P_\Lambda) = \min_P \sum_k \lambda_k D\xi_k(P)$$

with respect to λ_i at the point $\Lambda = (\lambda_1^{(0)}, \ldots, \lambda_n^{(0)})$ using the relation $\lambda_{i_0} = 1 - \sum_i' \lambda_i$. Here $\sum'$ means that the term λ_{i_0} is omitted. Thus,

$$\frac{\partial G(\Lambda_0)}{\partial \lambda_i} = D\xi_i(P_{\Lambda_0}) - D\xi_{i_0}(P_{\Lambda_0}) + \sum_k \lambda_k^{(0)} \frac{\partial \xi_k(P_{\Lambda_0})}{\partial \lambda_i}. \tag{4.4}$$

If $0 < \lambda_i^{(0)} < 1$, then (4.4) vanishes, thus we get from (4.3)

$$D\xi_i(P_{\Lambda_0}) = D\xi_{i_0}(P_{\Lambda_0}).$$

If $\lambda_i^{(0)} = 0$, then (4.4) is nonpositive, thus from (4.3) we get

$$D\xi_i(P_{\Lambda_0}) \le D\xi_{i_0}(P_{\Lambda_0}).$$

Notice that $\lambda_i^{(0)} = 0$ for $i \ne i_0$ if $\lambda_{i_0}^{(0)} = 1$. However the last inequality remains true in this case also.

Thus, the variances are equal on the support of the measure of Λ_0, while the others do not exeed this value. It remains to apply the standard inequality [4.2].

From the derived relations we get

$$F_\xi(P_{\Lambda_0}) = \sum_i \lambda_i^{(0)} D\xi_i(\Lambda_0) = G(\Lambda_0).$$

On the other hand,

$$F_\xi(P) \geq \sum_i \lambda_i^{(0)} D\xi_i(P) \geq \sum_i \lambda_i^{(0)} D\xi(P_{\Lambda_0}) = G(\Lambda_0). \qquad \square$$

It is clear that it may appear that the quantities $D\xi_k(P)$ are not the variances of ξ_k. The notation $D\xi_k$ is used here for simplicity only.

Notice that Lemma 4.1 cannot be derived directly from [4.2], since in [4.2] it is assumed that the set $\{P\}$ is compact. Formally, it can be considered as a special case of the minimax theorem [4.3], since

$$\max_k D\xi_k = \max_\Lambda \left\{ \sum_{k=1}^n \lambda_k D\xi_k | \sum_k \lambda_k = 1, \lambda_k \geq 0 \right\}.$$

To complete the proof it is sufficient to invert the order of min and max. However, in the standard minimax theorem the set $\{P\}$ is also assumed to be compact. Therefore, the simple proof of Lemma 4.1 presented is reasonable.

4.2 The Minimax Estimates for the Integrals

Note that the solution to the problem of optimization of estimates for integrals according to the criterion based on the weighted sum of variances is well-known (Sect. 1.5) and is given by

$$\min \sum_{k=1}^n \lambda_k D\zeta_k(p) = \sum_{k=1}^n \lambda_k D\zeta_k(p_\Lambda) = G_n(\Lambda),$$

where

$$p_\Lambda(x) = c\left[\sum_{k=1}^n \lambda_k g_k^2(x)\right]^{1/2}, \quad c^{-1} = \int_X \left[\sum_{k=1}^n \lambda_k g_k^2(x)\right]^{1/2} dx,$$

$$G_n(\Lambda) = \left\{ \int_X \left[\sum_{k=1}^n \lambda_k g_k^2(x)\right]^{1/2} dx \right\}^2 - \sum_{k=1}^n \lambda_k I_k^2.$$

Here, the nature of the space X and of the integration measure is not important. However, we suppose for simplicity that X is a finite-dimensional Euclidian space, and the integrals are taken with respect to the Lebesgue measure in X. Below, we consider the statement concerning the minimax of the variances of estimates of n integrals.

Theorem 4.1. Let

$$P_0 = \int_X \sup_k |g_k(x)|\, dx < \infty.$$

Then the optimal density is given by $p^*(x) = p_{\Lambda^*}(x)$ where Λ^* is a maximum point for $G_n(\Lambda)$.

Proof. The condition of finiteness of (4.1) is satisfied since

$$G_n = G_n(\Lambda) \leq 2\left(\sum_{k=1}^{n} \lambda_k\right) P_0^2.$$

It is sufficient now to verify that the functions

$$\mathbf{M}\zeta_m^2(\Lambda) = \int_X \left[\sum_{k=1}^{n} \lambda_k g_k^2(x)\right]^{1/2} dx \int_X g_m^2(x) \left[\sum_{k=1}^{n} \lambda_k g_k^2(x)\right]^{-1/2} dx$$
$$= S_m(\Lambda) R_m(\Lambda).$$

are differentiable. Differentiating S_m with respect to λ_i yields

$$S_m^{(i)}(\Lambda) = \frac{1}{2} \int_X g_i^2(x) \left[\sum_{k=1}^{n} \lambda_k g_k^2(x)\right]^{-1/2} dx.$$

The integrand is a monotonic function of λ_i for $\lambda_i \geq 0$; therefore we use the rule

$$[S_m^(\Lambda)]'_{\lambda_i} = S_m^{(i)}(\Lambda).$$

Differentiability of the function $R_m(\Lambda)$ is derived analogously. □

We also consider the case where $g_0 = g(x, \sigma)$ depends continuously on σ, $\sigma \in [\sigma^{(1)}, \sigma^{(2)}]$. We introduce the following notation: Λ is a probabilistic measure on $[\sigma^{(1)}, \sigma^{(2)}]$, $\zeta_\sigma(p) = \zeta(p, \sigma)$,

$$P_0 = \int_X [\sup_\sigma |(x, \sigma)|] \, dx,$$

$$\varphi^2(x) = \inf_k \int_X g^2(x, \sigma) \Lambda_k^*(d\sigma),$$

where $\Lambda_k^* = (\lambda_1^*, \ldots, \lambda_{k+1}^*)$ is the optimal (in the sense of Theorem 4.1) discrete measure for the set of values of σ coinciding with the k nodes of the uniform network of $[\sigma^{(1)}, \sigma^{(2)}]$.

Theorem 4.2. Let

$$P_0 < \infty, \quad |g(x, \sigma) g'_\sigma(x, \sigma)| \leq h(x),$$

$$\int_X \frac{g^2(x, \sigma)}{\varphi(x)} \, dx < c_1 < \infty, \quad \int_X \frac{h(x)}{\varphi(x)} \, dx < c_2 < \infty. \tag{4.5}$$

Then

$$p^*(x) = p(x, \Lambda^*) = c \left[\int_X g^2(x,\sigma)\Lambda^*(d\sigma) \right]^{1/2},$$

where Λ^* is a maximum point for $G(\Lambda)$:

$$G(\Lambda) = \left\{ \int_X \left[\int_X g^2(x,\sigma)\Lambda^*(d\sigma) \right]^{1/2} dx \right\}^2 - \int_X I^2(\sigma)\Lambda(d\sigma).$$

Proof. Let $\sigma_1^{(n)}, \ldots, \sigma_n^{(n)}$ be the nodes of the uniform subdivision of the segment $[\sigma^{(1)}, \sigma^{(2)}]$ into $n - 1 = 2^m$ parts, and let $g_{in}(x) = g(x, \sigma_i^{(n)})$. Then

$$G_n^* = G_n(\lambda_{in}^*, \ldots, \lambda_{nn}^*) \uparrow G^n \tag{4.6}$$

as $n \to \infty$.

The set of probabilistic measures on a finite segment is relatively compact [4.4]. Therefore, from the sequence of discrete measures

$$\{(\sigma_1^{(n)}, \lambda_{1n}^*), \ldots, (\sigma_n^{(n)}, \lambda_{nn}^*)\}, \quad n = 1, 2, \ldots$$

it is possible to extract a subsequence of measures Λ_k^* which is weakly convergent to a probablistic measure Λ^*. Thus from (4.6) we obtain $G_n^* \uparrow G(\Lambda^*) = G^*$. Next, the conditions (4.5) ensure that

$$\begin{aligned} F(p^*) &= \sup_\sigma D\zeta(p^*, \sigma) \\ &= \sup_\sigma \left[\int \frac{g^2(x,\sigma)}{p^*(x)}\, dx - I^2(\sigma) \right] = \lim_{k\to\infty} F(p_k^*) \\ &= \lim \sup_\sigma \left[\int \frac{g^2(x,\sigma)}{p_k^*(x)}\, dx - I^2(\sigma) \right], \end{aligned} \tag{4.7}$$

since

$$g^2(x,\sigma) = \left[\int g^2(x,\sigma)\Lambda_k^*(d\sigma) \right]^{1/2} \le \frac{g^2(x,\sigma)}{\varphi(x)}.$$

To compare $F(p_k^*)$ and G_k^* we note that

$$G_k^* = \sup_\sigma \int_X \frac{g_1^2(x,\sigma)}{p_k^*(x)}\, dx,$$

where

$$g_1(x,\sigma) = g(x, \sigma_i^{(k)}), \quad \sigma_i^{(k)} < \sigma < \sigma_{i+1}^{(k)}, \quad i = 1, \ldots, n_k - 1,$$

$$g_1(x, \sigma_i^{(k)}) = g(x, \sigma_i^{(k)}), \quad i = 1, \ldots, n_k.$$

Therefore,

$$|F(p_k^*) - G_k^*| \le P_0 c^2 O(n_k^{-1}). \tag{4.8}$$

From (4.6, 8) it follows that $F(p^*) = G(\Lambda^*)$. Then, applying the arguments used in the second part of the proof of Lemma 4.1, we conclude that p^* is optimal. The measure Λ^* is the maximum point of $G(\Lambda)$, since

$$F(p^*) \ge \int D\zeta(p^*, \sigma)\Lambda(d\sigma) \ge G(\Lambda) \quad \forall\, \Lambda. \qquad \square$$

The conditions of this theorem are essentially simple if $g(x, \sigma)$ is defined on a rectangle and separated from zero. The theorem is automatically generalized to the case of integrals over an arbitrary measure $P(dx)$ and when λ is multidimensional. The theorem remains valid if the interval $(\sigma^{(1)}, \sigma^{(2)})$ is replaced with a bounded domain which consists of segments and points. It is possible to consider an approximate discrete problem. Using the relation $F(p) \ge G^* \ge G_n^*$, we can control the approximate solution: if for some $p(x)$ and n we obtain $F(p) \gtrsim G_n^*$, then $p(x)$ is approximately optimal.

A series of examples has shown that sometimes it is useful to take

$$p \equiv p_0(x) = P_0^{-1} \sup_\sigma |g(x, \sigma)| \Rightarrow D\zeta(p_0, \sigma) \le I_1(\sigma)[P_0 - I^2(\sigma)/I_1].$$

If the relative variances $D\zeta(p, \sigma)/I^2(\sigma)$ are to be minimized, then $g(x, \sigma)$ is replaced by

$$\frac{g(x, \sigma)|}{I_1(\sigma)}, \quad I_1(\sigma) = \int_X |g(x, \sigma)|\, dx.$$

If two integrals of nonnegative functions g_1, g_2 are calculated, it is then useful to take the density

$$p^{(0)}(x) = \frac{g_1(x) + g_2(x)}{I_1 + I_2}.$$

Now,

$$\zeta_2 = (\zeta_1 + \zeta_2) - \zeta_1, \quad D[\zeta_1(p^{(0)}) + \zeta_2(p^{(0)})] = 0,$$

therefore $D\zeta_1(p^{(0)}) = D\zeta_2(p^{(0)})$. Consequently, if $g_1 g_2 \equiv 0$, then $p^* = p^{(0)}$, since in this case $g_1^2 + g_2^2 \equiv (g_1 + g_2)^2$. On the other hand, $p^{(0)}$ is optimal if $g_1 \equiv g_2$. So it is hoped that $p^{(0)}$ will give a good result in a broad class of cases (Sect. 4.3). It is not difficult to see that the relative variances are equal if $p_1^{(0)} = g_1/I_1 + g_2/I_2$.

4.3 Optimization of Estimates for the Integral Equations

Assume that it is desirable to calculate a linear functional $I = (\varphi^*, h)$, where $\varphi^* = K^*\varphi^* + f$, and K^* is an integral operator with a kernel $k(x', x)$. It is assumed that $\rho(K_1^*) < 1$, where K_1^* is the operator with the kernel $|k(x', x)|$. Define a terminating Markov chain to be $P = \{x_n\}$ with the initial density $\pi(x)$ and the transition density $p(x', x) = p(x' \to x)$. The quantity $g(x') = 1 - \int p(x'x)\, dx$ is interpreted as a terminating probability at the point x'; N is the random number of the last state. The collision estimate ξ and the absorption estimate η are defined by (Sect. 1.6)

$$\xi = \sum_{n=0}^{N} Q_n h(x_n), \quad \eta = \frac{Q_N h(x_N)}{g(x_N)},$$

where

$$Q_0 = f(x_0)/\pi(x_0),$$

$$Q_n = Q_{n-1}\, k(x_{n-1}, x_n)/p(x_{n-1}, x_n), \quad n \geq 1.$$

If $p(x', x) \neq 0$ when $k(x', x) \neq 0$ and $\pi(x) \neq 0$ when $f(x) \neq 0$, then $\mathbf{M}\xi = (\varphi^*, h)$. If, in addition, $g(x) \neq 0$ when $h(x) \neq 0$, then $\mathbf{M}\eta = (\varphi^*, h)$ (Sect. 2.2). For simplicity, we assume that the functions mentioned above are all positive.

The results of Sect. 4.2 can be directly applied to the uniform optimization of the estimates η_σ related to the functions $h_\sigma = h(x, \sigma)$, since the estimate η with zero variance corresponds to the importance sampling for the integral representing $\mathbf{M}\eta$ (Sect. 2.1). Here the optimal Markov chain P_{Λ^*} is defined by the importance function

$$\varphi = K\varphi + h^*, \quad h^* = \left[\int h_\sigma^2 \Lambda^*(d\sigma)\right]^{1/2},$$

where Λ^* is the solution to the problem

$$\max_{\Lambda} \left\{ \left(\varphi^*, \left[\int h_\sigma^2 \Lambda^*(d\sigma)\right]^{1/2}\right)^2 - \int (\varphi^*, h)^2 \Lambda(d\sigma) \right\}.$$

For this chain

$$p(x', x) = k(x', x)\varphi(x)/\varphi(x'),$$

$$\pi(x) = f(x)\varphi(x)/(\varphi, h^*).$$

Let us optimize the collision estimate

$$\xi_k = \xi(P, \sigma_k), \quad k = 1, \ldots, n.$$

Note that there exists a unique (Sect. 2.1) chain $P_\Lambda^{(0)}$, $\Lambda = (\lambda_1, \ldots, \lambda_n)$, such that $\min_P \sum_k \lambda_k D\xi_k$ is attained. Thus, Lemma 4.1 is applicable, i.e., the minimum

$$\min \max_k D\xi(P, \sigma_k)$$

is attained for the chain $P^{(0)}_{\Lambda_0}$, where $\Lambda = (\lambda_1^{(0)}, \dots, \lambda_n^{(0)})$ is the solution to the problem

$$\max \Lambda \left[\sum_{k=1}^{n} \lambda_k D\xi(P^{(0)}_{\Lambda}, \sigma_k) \mid \sum_{k=1}^{n} \lambda_k = 1 \right] \tag{4.9}$$

Theorem 4.3. Let

$$D^* = \min_P \max_k D\eta(P, \sigma_k), \qquad D^{(0)} = \min_P \max_k D\xi(P, \sigma_k).$$

Then $D^{(0)} \leq D^*$.

Proof. It is known (Sect. 2.1) that

$$\sum_{k=1}^{n} \lambda_k D\xi(P^{(0)}_{\Lambda}, \sigma_k) \leq \sum_{k=1}^{n} \lambda_k D\eta(P_{\Lambda}, \sigma_k).$$

Therefore it follows from Lemma 4.1 that

$$D^{(0)} = \sum_{k=1}^{n} \lambda_k^{(0)} D\xi(P^{(0)}_{\Lambda_0}, \sigma_k) \leq \sum_{k=1}^{n} \lambda_k^{(0)} D\eta(P_{\Lambda_0}, \sigma_k) \leq D^*. \qquad \square$$

When $n = 2$, the measure Λ is defined by a real number λ; thus the optimization of the estimates ξ and η reduces to solving the equations

$$D\xi(P^{(0)}_{\Lambda}, \sigma_1) = D\xi(P^{(0)}_{\Lambda}, \sigma_2),$$

$$D\eta(P_{\Lambda}, \sigma_1) = D\eta(P_{\Lambda}, \sigma_2). \tag{4.10}$$

If $h_1 h_2 \equiv 0$, then $h^* = c(h_1 + h_2)$, since in this case the second equation is satisfied. This is also true for the estimates ξ_1 and ξ_2 which are optimal in the class of chains P_{Λ} for $g \equiv 0$.

To compare the absorption and collision estimates we give the following illustrative example. The components of the system

$$x = ax + by + 1,$$

$$y = bx + ay + 1$$

were calculated by the Monte Carlo method [4.1]. Here $a, b \geq 0$, $a + b = q < 1$. Since $h_1\, h_2 \equiv 0$, the estimates η_1, η_2 are optimal if $D(\eta_1 + \eta_2) = 0$. Due to the symmmetry of the problem the same Markov chain but with a zero terminating probability gives the best ξ_1, ξ_2. Using standard formulae (Sect. 1.6) we obtained the following expressions for the variances:

$$D\eta_k^* = (1 - q)^2,$$

$$D\xi_k^{(0)} = (1-b+a)/[(1-q^2)(1+b-a)], \quad k=1,2.$$

Dividing the first expression by the second, we obtain

$$r(a,b) = (1+q)(1+b-a)/[(1-q)(1-b+a)] \geq 1.$$

For example, $r(0.6, 0.2) = 21$, $r(0.2, 0.6) = 27/7$.

We now consider an important problem for constructing weighted estimates which are effective for calculating the functionals of the solutions of a set of integral equations by simulating the same Markov chain. In principle, the weighted estimates can be interpreted as vector estimates; thus the solution to the problem can be written as in (4.9), since there exists a unique chain for which the minimum in (4.9) is attained (Sect. 5.9). However this chain is non-Markov and there are difficulties in simulating such chains. Therefore we use a special approximate solution. Additional arguments which show some advantages of this approximation will be given in Sects. 4.1, 2. The mean squares of the collision and absorption estimates can be written in the form (χ, q_h), where $q_h(x)$ are the weight functions related to $h(x)$, and χ is defined by (Sect. 1.6)

$$\chi(x) = \int \frac{k^2(x', x)}{p(x', x)} \chi(x')\, dx' + \frac{f^2(x)}{\pi(x)}.$$

This equation shows that in order to optimize the estimates uniformly, it is necessary to take $p(x', x)$ such that the quantity

$$\sup_x \int \frac{k^2(x', x)}{p(x', x)}\, dx$$

is minimal. If this quantity is less than unity then the collision estimates have finite variances.

If the solution depends on some parameter σ, then it is necessary to minimize the following quantity

$$R(p) = \sup_\sigma \sup_{x'} \int \frac{k^2(x', x, \sigma)}{p(x', x)} dx = \sup_{x'} \sup_\sigma \int \frac{k^2}{p}\, dx.$$

This can be done by solving

$$\inf_p \sup_\sigma \int \frac{k^2(x', x, \sigma)}{p(x', x)}\, dx \quad \forall\, x' \in X. \tag{4.11}$$

The solution is based on Theorem 4.2, where $D\zeta_\sigma$ is replaced with $\mathbf{M}\zeta_0^2$, i.e., for

$$G(\Lambda) = \left\{ \int \left[\int k^2(x', x, \sigma)\, \Lambda(d\sigma) \right]^{1/2} dx \right\}^2.$$

Let us now consider a radiative transfer problem which can be solved by the method of similar trajectories (Sect. 1.6). The integral transfer equation is solved for a medium with a certain geometry but with different total cross-sections $\sigma \in (\sigma_1, \sigma_2)$ by simulating one and the same trajectory with the transition density $p(x, x')$. For simplicity we assume that the bounding surface is convex and the space outside of the domain is filled with an absolute absorber; thus, the kernel is normalized with respect to the spatial variable (Sect. 1.3). In addition, it is assumed that the direct simulation of the scattering is carried out, i.e., the angular factor of the density $p(x', x)$ is equal to that of the kernel $k(x', x, \sigma)$. This factor is omitted after the integration with respect to the scattering angle. Therefore it is possible to take

$$k_0(x', x, \sigma) = \sigma \exp(-\sigma t), \quad t > 0,$$

where t is the free length between x' and x. Consequently, the minimization of $R(p)$ reduces here to the construction of the density $p^*(t)$ such that $\min F_\xi(p)$ is attained, where

$$F_\xi(p) = \max_{\sigma_1 \leq \sigma \leq \sigma_2} \int \frac{\sigma^2 \exp\{-2\sigma t\}}{p(t)} \, dt.$$

Notice that the method of similar trajectories consists in applying the densities of the type $p_1^*(t) = \sigma^* \exp(-\sigma t), (t > 0)$. Simple arguments show that the optimal value is $\sigma^* = 2\sigma_1\sigma_2/(\sigma_1 + \sigma_2)$, and

$$F_\xi(p_1^*) = 1 + \frac{1}{4}\left(\frac{\sigma_1}{\sigma_2} + \frac{\sigma_2}{\sigma_1} - 2\right).$$

It was found that the optimal density can be written with high accuracy as

$$p^*(t) = [\lambda\sigma_1^2 \exp\{-2\sigma_1 t\} + (1-\lambda)\sigma_2^2 \exp\{-2\sigma_2 t\}]^{1/2},$$

i.e., the measure $\Lambda^*(d\sigma)$ is concentrated on the end points of the interval (σ_1, σ_2). This follows from (Lemma 4.1)

$$\mathbf{M}\xi^2(\sigma_1, p^*) = \mathbf{M}\xi^2(\sigma_2, p^*)$$

$$\mathbf{M}\xi^2(\sigma, p^*) \leq \mathbf{M}\xi^2(\sigma_1, p^*), \quad \sigma_1 \leq \sigma \leq \sigma_2.$$

These relations were verified numerically with high accuracy for a wide range of values of σ_2 for $\sigma_1 = 1$. The calculations show that $F_\zeta(p^*) = 1.54$ for $\sigma_1 = 1, \sigma_2 = 10$ while $F_\zeta(p_1^*) = 3.525$. However, the density $p^*(t)$ is not convenient in practical applications of the weighted algorithms to the transfer problems. Therefore, it is necessary to use simpler approximations of the optimal densities. The minimax density p^* also solves the minimax problem with two values of the parameter; thus it is useful to take (Sect. 4.2)

$$p^{(0)}(t) = (\sigma_1 \exp(-\sigma_1 t) + \sigma_2 \exp(-\sigma_2 t))/2.$$

The calculations show that $D\xi(\sigma, p^{(0)}) \leq D\xi(\sigma_1, p^{(0)})$, $(\sigma_1 \leq \sigma \leq \sigma_2)$ and $F_\xi(p^{(0)}) = D\xi(\sigma_1, p^{(0)}) = 1.56$ $(\sigma_1 = 1,\ \sigma_2 = 10)$, i.e., the density $p^{(0)}(t)$ is effective.

Note that it is not necessary to verify the conditions of Theorem 4.2, since the conditions of Theorem 4.1 for two integrals are satisfied and the following inequalities hold: $G^* \geq G_2^*$, $F(p^{(0)}) \geq G^*$, i.e $1.54 \leq G^* \leq 1.56$, where $G^* = \min F(p)$.

For solving systems of integral equations, it is possible to use the vector Monte Carlo algorithms based on using the matrix weight which after the transition $x' \to x$ is multiplied from the left by $K(x',x)/p(x'x)$, where $K(x',x)$ is the matrix of the kernels of the system under study. The covariance matrix of the vector estimate is defined by a special matrix-integral equation. The norm of the corresponding operator is defined by [4.5]

$$\sup_{i,x'} \int \left[\sum_{j=1}^{m} |k_{ij}(x',x)| \right]^2 [p(x',x)]^{-1} \, dx.$$

Thus, to minimize the vector estimates uniformly, it is useful to solve the minimax problem of the type (4.10) by the method described in Theorem 4.1. This approach will be described in detail in Chap. 5.

4.4 Minimax Choice of the First Step in the Markov Chain

Let us consider the standard estimate for solving the integral equation $\varphi = K\varphi + h$ in the form

$$\xi_x = h(x) + \frac{k(x,x_1)}{p(x,x_1)} \, \xi_{x_1}.$$

We now assume that the first step $x \to x_1$ is simulated according to the density $p_0(x_1)$ which is independent of x and such that $p_0(x_1) \neq 0$ if $k(x,x_1) \neq 0$ for all $x \in D \subset X$. Then the random estimate

$$\xi_x^{(0)} = h(x) + \frac{k(x,x_1)}{p_0(x_1)} \, \xi_{x_1} \tag{4.12}$$

is an unbiased estimate for the solution $\varphi(x)$, i.e., $\mathrm{M}\xi_x^{(0)} = \varphi(x)$ for arbitrary $x \in X$. Consequently, by applying (4.12), it is possible to calculate the solution at a set of points in D using the same set of trajectories $\{x_n\}$ $(n = 1, 2 \ldots)$.

The variance of $\xi_x^{(0)}$ is given by the formula

$$D\xi_x^{(0)} = \int_X \frac{k^2(x,x_1)}{p_0(x_1)} \, \mathrm{M}\xi_{x_1}^2 \, dx_1 - \left[\int_X k(x,x_1)\varphi(x_1)dx_1 \right]^2.$$

Let us now consider the problem of finding the density $p_0^*(x_1)$ such that

$$\min_{p_0} \max_x D\xi_x^{(0)} \tag{4.13}$$

is attained. For simplicity we suppose that the set D consists of a finite number of points. Assume also that

$$\int_X \sup_x \left[|k(x,x_1)| \sqrt{\mathrm{M}\xi_{x_1}^2} \right] dx_1 < \infty.$$

Then Lemma 4.1 and Theorem 4.1 are applicable and

$$p_0^*(x_1) = p_0^*(x_1, \Lambda^*) = \frac{[\mathrm{M}\xi_{x_1}^2 \int_X k^2(x,x_1)\Lambda^*(dx)]^{1/2}}{\int_X [\mathrm{M}\xi_{x_1}^2 \int\_D k^2(x,x_1)\Lambda^*(dx)]^{1/2}\, dx_1},$$

where Λ^* is the maximum point of the function

$$G(\Lambda) = \left\{ \int_X \left[\mathrm{M}\xi_{x_1}^2 \int_D k^2(x,x_1)\Lambda^*(dx) \right]^{1/2} dx_1 \right\}^2$$
$$- \int_X \int_D k(x,x_1)\varphi(x_1)]^2 \Lambda(dx).$$

Here Λ is the probabilistic measure on D.

The expression for $p_0^*(x)$ is simplified when we use the approximation $\mathrm{M}\xi_{x_1}^2 \approx$const which makes sense if the function

$$\int_D k^2(x,x_1)\Lambda^*(dx)$$

varies more strongly than $\mathrm{M}\xi_{x_1}^2$. Sometimes, the approximation $\mathrm{M}\xi_{x_1}^2 \approx c\varphi^2(x)$ is preferable.

As a special case of (4.12), the known estimate for solving the Poisson equation (Sects 1.4, 6) can be considered. Thus, the optimization technique just described can be applied. The kernel $k(x,x_1)$ is then the normal derivative of Green's function at the points of the sphere $S(x_0)$, and the integration with respect to x_1 is carried out over $S(x_0)$. The function $h(x)$ is represented through the integral of the product of Green's function and the right-hand side of the Poison equation over the ball bounded by $S(x_0)$. If this integral is calculated by the Monte Carlo method using one random node, then the density of this node can be constructed on the basis of the minimax approach described in Sect. 4.2. Note that if the approximation $\mathrm{M}\xi_{x_1}^2 \approx$const is used and the domain D has an axial symmetry (with respect to x_0), then the minimax distribution of x_1 is uniform over $S(x_0)$. This follows from the symmetry of Green's function.

5. Vector Monte Carlo Algorithms

Vector Monte Carlo algorithms are used for solving systems of integral equations. The distinguishing feature of the vector algorithm is that its "weight" appears to be a matrix weight, which is multiplied by the kernel matrix of the system of integral equations divided by a transition density function after each transition in the Markov chain simulation. The following vector Monte Carlo algorithms are well-known: an algorithm for solving the system of transfer equations, considering polarization [5.1-3], a vector algorithm for solving multi-group transfer eauations [5.4], and a Monte Carlo method for solving metaharmonic equations [5.5]. We also mention the weighted algorithms, so-called depending sampling methods [5.6], which are used in solving equations whose kernel contains a parameter. A matrix-integral equation which determines the variance of the vector estimates was obtained in [5.7].

We will consider applications of this method for improving algorithms for solving multigroup transfer equations, and also the Monte Carlo technique combined with the finite sum method. The variance of the estimates for transfer equations considering polarization is investigated. Vector Monte Carlo algorithms for solving systems of integral equations with a triangular matrix of kernels are also presented. A simple variance finiteness condition for the vector algorithm proposed in [5.4] for solving multigroup transfer equations is formulated. Systems of integral equations for the iterative solution of integral equations of the second kind where the Neumann series diverges are proposed. Vector weighted estimates which can be used for finding the bilinear functionals of perturbation theory are constructed.

The problem of variance finiteness for the estimates of derivatives of the linear functionals with respect to parameters is considered. A new vector Monte Carlo algorithm is given for evaluating K_{eff}, the effective multiplication coefficient.

5.1 Variance Vector Algorithms

We first consider a system of integral equations of the second kind,i.e.,

$$\varphi_i(x) = \sum_{j=1}^{m} \int_X k_{ij}(x,y)\varphi_j(y)dy + h_i(x), \tag{5.1}$$

or in the operator form

$$\Phi = K\Phi + H,$$

where $H' = (h_1, ..., h_m) \in L_\infty, K \in L_\infty \to L_\infty, \Phi' = (\varphi_1, ..., \varphi_m)$,

$$\|H\|_{L_\infty} = \operatorname{vrai}\sup_{i,x} |h_j(x)|;$$

and the integration is performed with respect to the Lebesgue measure in the Euclidean space X.

It is supposed that $\rho(K) < 1$, where $\rho(K)$ is the spectral radius of K. Let us also consider a Markov chain $\{x_n\}, n = 0, 1, ..., N$ with the transition density function $p(x, y)$, where the quantity $1 - \int p(x, y)dy \geq 0$ is treated as the probability of the chain break (absorption state) at a point x, N is a random number of the last state and $x_0 = x$. The standard vector Monte Carlo estimate for $\Phi(x)$ takes the form

$$\Phi(x) = \mathbf{M}\xi_x, \quad \xi_x = H(x) + \sum_{n=1}^{N} Q_n H(x_n),$$

$$Q_0 = I, \quad Q_{n+1} = Q_n K(x_n, x_{n+1})/p(x_n, x_{n+1}), \quad n = 0, 1, ... \; .$$

$Q_0 = J$ is the identity matrix, and $K(x, y)$ is a matrix $\{k_{ij}(x, y)\}, i, j = 1, ..., n$. It should be noted that $\Phi(x) = \mathbf{M}\xi_x$ holds under well-known conditions of unbiasedness and if $\rho(K_1) < 1$, where the operator K_1 is obtained from K by replacing k_{ij} with $|k_{ij}|$ [5.8]. In [5.8], the equation for the covariance martix $\Psi(x) = \mathbf{M}(\xi_x \xi_x^T)$ was obtained by

$$\Psi(x) = \chi(\lambda) + \int_X \frac{K(x, y)\Psi(y)K'(x, y)}{p(x, y)} dy, \tag{5.2}$$

where $\chi = H\Phi' + \Phi H' - HH'$. Equation (5.2) is considered in the space L_∞ of matrix-valued functions with a norm

$$\|\varphi\| = \operatorname{vrai}\sup_{x,i,j} |\varphi_{i,j}(x)|.$$

Hereafter the symbol vrai will be omitted.

A simple method of the derivation of (5.2) will now be considered. This method is used in the next section. We also obtain an expression for the norm of the matrix-integral operator $K_p \in [L_\infty \to L_\infty]$, defined by (5.2). Let K_{p1} be a matrix-integral operator obtained from K_p by replacing the kernel k_{ij} with $|k_{ij}|$.

Lemma 5.1. If $\rho(K_1) < 1$, and $\rho(K_{p1}) < 1$, then $\Psi(x) = \mathbf{M}(\xi_x \xi_x')$ solves (1.2).

Proof. From the definition of ξ_x it follows that

$$\xi_x \xi'_x = \left[H(x) + \frac{K(x,x_1)}{p(x,x_1)}\xi_{x_1}\right]\left[H'(x) + \xi'_{x_1}\frac{K'(x,x_1)}{p(x,x_1)}\right]$$
$$= HH' + \frac{K(x,x')}{p(x,x')}\xi_{x_1}H' + H\xi'_{x_1}\frac{K'(x,x_1)}{p(x,x_1)}$$
$$+ \frac{K(x,x_1)}{p(x,x_1)}\xi_{x_1}\xi'_{x_1}\frac{K'(x,x_1)}{p(x,x_1)},$$

where $H = H(x)$. The mathematical expectations of the second and third terms are equal to $(\Phi - H)H'$ and $H(\Phi' - H')$, respectively. Their sum when added to HH' gives χ. The mathematical expectation of the absolute value of the fourth term is finite since $\rho(K_{p,1}) < 1$. Therefore, this expectation can be performed as follows: one takes the condition expectation under the condition that x_1 is fixed and then averages the result over x_1, which gives the integral in (5.2). □

Lemma 5.2. Let the kernels of (5.1) be measurable functions of the variables (x, y) in the spase $X \times X$. Then

$$\|K_p\| = \sup_{i,x} \int \frac{\left(\sum_{j=1}^{m} |k_{ij}(x,y)|\right)^2}{p(x,y)} dy \quad . \tag{5.3}$$

Proof. The operator K_p is a linear integral operator defined on the vector functions Φ with components Φ_{ij}. By the definition

$$\|K_p\| = \|K_{p1}\| = \left\| \int \frac{K_1(x,y)LK'_1(x,y)}{p(x,y)} dy \right\|,$$

where K_1 is a matrix whose entries are the absolute values of the kernels and L is a matrix with unit entries: $L_{ij} \equiv 1$. We can write $L = ll'$, where $l' = (1, ..., 1)$; hence $K_1LK'_1 = (K_1l)(K_1l)'$,

$$(K_1LK'_1)_{it} = \left(\sum_{j=1}^{m} |k_{ij}(x,y)|\right)\left(\sum_{j=1}^{m} |k_{tj}(x,y)|\right).$$

From the Cauchy-Schwarz-Buniakowski inequality it follows that the non-diagonal elements of $K_{p1}L$ are majorized by the diagonal one; hence Lemma 5.2 is proved. □

By the Monte Carlo methods one can calculate the linear functionals of the type

$$I = (F, \Phi) = \int F'(x)\Phi(x)dx,$$

where $f'(x) = (f_1(x), ..., f_m(x))$,

$$\|F\|_{L_1} = \sum_j \int |f_j(x)|dx < \infty.$$

Assume that the point is distributed according to the probability density $\pi(x)$ such that $\pi(x) \neq 0$, when $F'(x)\Phi(x) \neq 0$. Then

$$I = \mathbf{M}\left\{\frac{F'(x_0)}{\pi(x_0)}\xi_{x_0}\right\}$$
$$= \mathbf{M}\left\{\sum_{n=0}^{N} \frac{f'(x_0)}{\pi(x_0)} Q_n H(x_n)\right\} = \mathbf{M}\left\{\sum_{n=0}^{N} H'(x_n)Q'_n \frac{F(x_0)}{\pi(x_0)}\right\}.$$

The random vector weights

$$Q_n^{(')} = Q_n^{(')} F(x_0)/\pi(x_0)$$

are calculated as follows:

$$Q_n^{(')} = [K'(x_{n-1}, x_n)/p(x_{n-1}, x_n)]Q_{n-1}^{(')}.$$

This vector algorithm corresponding to the representation $I = (\Phi^*, H)$ is formulated in [5.3]. Note that in [5.3], $\Phi^* = K^*\Phi^* + F$ is considered to be a basic equation, and the weights $Q_n^{(')}$ are transformed by a nonconjugate matrix kernel. Let

$$\zeta = F'(x_0)\xi_{x_0}/\pi(x_0),$$

then

$$\mathbf{M}\zeta^2 = \mathbf{M}\{F'(x_0)\xi_{x_0}\xi' x_0 F(x_0)/\pi^2(x_0)\}$$
$$= \mathbf{M}\{F'(x_0)\Psi(x_0)F(x_0)/\pi^2(x_0)\}.$$

Thus the variance $\mathbf{D}\zeta$ is determined by the matrix valued function $\Psi(x)$. In particular, if $\rho(K_p) < 1$, $|F(x)/\pi(x)| < \delta < \infty$.

The cone of covariance matrix functions is invariant under the transformation K_p. Therefore, a natural assumption can be made that there exist a maximal eigenvalue of K_p and the corresponding eigen-covariance matrix $S(x)$ (e.g., when K_p is compact). In what follows the inequality

$$\alpha[1 - \|k_p\|]^{-1} \leq \sup_{\|x\|=1} \|\Phi\| \leq \|K_p\|]^{-1}. \tag{5.4}$$

will be used. We could expect that, in some cases, the constant α in (5.4) does not differ much from 1. This conjecture follows, for example, from the following statement, where the matrix $K(x, y)$ is assumed to be diagonal (as in the depending sampling technique).

Theorem 5.1. Assume that $k_{ij} \equiv 0$ for $i \neq j$, that there exist a maximal eigenvalue of K_p and the corresponding eigen-covariance matrix $S(x)$, and that $\|s\| = 1, \|K_p\| < 1$. Then

$$[1-\rho(K_p)]^{-1} \leq \sup_{\|\chi\|=1} \|\Phi\| \leq \|K_p\|]^{-1}. \tag{5.5}$$

Proof. The right-hand side of (5.5) is obvious, and the left-hand side is true if H exists such that $\chi = S$. Indeed, in this case the norm of the solution of (5.2) is equal to $[1-\rho(K_p)]^{-1}$. Now, the equation $\chi = s$ can be written as

$$h_i(x)[2\varphi_i(x) - h_i(x)] = S_{ii},$$

$$\varphi_i = K_{ii}\varphi_i + h_i, \quad i = 1, ..., m,$$

which is resolvable (Theorem 2.3). Theorem 5.1 is proved. □

Note that $\rho(K_p) = \|K_p\|$ if $K(x,y)$ and $P(x,y)$ depend only on $x - y$. Theorem 5.1 shows that in this case the bounds in (5.4) are, indeed, close under the condition that the last integral in (5.3) plays an essential role (as in the multigroup models of the transfer theory). Thus an optimal Markov chain for constructing uniform optimal weighted methods [i.e., optimal $p(x,y)$] can be obtained by solving a minimax problem (Chap.4):

$$\inf_p \sup_i \int \frac{1}{p(x,y)} \left(\sum_{j=1}^{m} |k_{ij}(x,y)| \right)^2 dy, \quad x \in X. \tag{5.6}$$

Note that the criterion (5.6) is somewhat stronger than the criterion $\inf \|K_p\|$. Optimization with respect to criterion (5.6) is uniform over x and therefore over linear functionals of the form (F', Φ) [5.8]. It should be noted that the numerical solutions of the transfer problems show that the variance maximum of the Monte Carlo estimates decreases or increases together with $\|K_p\|$ when $p(x,y)$ varies. This confirms the effectiveness of the criterion (5.6). Note that this criterion is exact when $N \equiv 1$ in the expression for ξ_x.

5.2 Uniform Optimization of Weighted Monte Carlo Estimates in the Transfer Theory

The weighted estimates are used, for example, for a simultaneous calculation of a set of integral equations with a kernel $k(x,y;\sigma)$, where σ is a parameter (e.g., the density of the medium, Sect. 1.6). The corresponding algorithm can be interpreted as a vector Monte Carlo estimate for solving systems with a diagonal matrix kernel. A simple investigation of the standard weighted Monte Carlo estimates for the transfer problems [5.7] shows that the lower and upper bounds of (5.4) are close if the optical depth τ is large.

Thus, the optimization of weighted estimates in the transfer theory can be carried out by solving the problem

$$\inf_p \sup_\sigma \int_X \frac{k^2(x,y;\sigma)}{p(x,y)} dy, \quad x \in X. \tag{5.7}$$

This problem is simplified if the kernel k depends only on the difference of the arguments; this is the case, e.g., if the medium is "almost homogeneous". Notice that the corresponding optimization of the method of "similar trajectories"(Sect. 1.6) is reduced to the following problem:

$$\min_{p} \max_{\sigma_1 \leq \sigma \leq \sigma_2} \int_0^\infty \frac{\sigma^2 \exp\{-2\sigma l\}}{p(l)} dl, \tag{5.8}$$

heuristically formulated in Sect. 4.4.

Numerical calculations show that for large τ a practically optimal density can be chosen as

$$p(l) = (\sigma_1 e^{-\sigma_1 l} + \sigma_2 e^{-\sigma_2 l})/2. \tag{5.9}$$

Notice that in the case of one equation the direct simulation technique is optimal as in (5.6).

If τ is small, it is usefull to take the density as in (5.9), where $\sigma e^{-\sigma l}$ is replaced with $\sigma e^{-\sigma l}(1 - e^{\sigma l_{\max}})^{-1}(0 \leq l \leq l_{\max})$. This means that the run without escape is simulated for both values of σ [5.7].

Let us now consider the problem of optimization of the vector algorithm for solving the multi-group system of transfer equations with an isotropic scattering [5.4]. In the homogeneous case we have (Sect. 1.3)

$$k_{ij}(x, y) = c\sigma_{ij} e^{-\sigma_i l},$$

where σ_i is the total transport cross section for i-th group, σ_{ij} is the cross section for the transition from i-th to j-th group, and l is the distance between the two phase points x and y. Thus (5.6) reduces to

$$\begin{aligned} &\min_{p} \max_{i} \int_0^\infty \frac{q_i^2 \sigma_i^2 \{-2\sigma_i l\}}{p(l)} dl, \\ &q_i = \sum_j \frac{\sigma_{ij}}{\sigma_i}, \end{aligned} \tag{5.10}$$

where $q_i < 1$ if the fission of particles is not taken into account.

A comparison of (5.8) and (5.10) shows that the effective optimal density takes the form of (5.9), where σ_1, σ_2 are replaced by the cross section of the groups which are close to the spectrum bounds of the system. This assumption was confirmed by numerical calculations carried out for a 26-group system for a critical ball of a fission medium [5.7]. The spatial fission density was $\sin(kr)/r$. The quantity (5.10) was calculated ($= 0.72$). The spectrum is concentrated in groups 0–11, and the effective optimal density is

$$p(l) = \lambda\sigma_2 e^{-\sigma_2 l} + (1 - \lambda)\sigma_{10} e^{-\sigma_{10} l}, \quad \lambda = 0.394. \tag{5.11}$$

We now make some comments on how these results were obtained. According to Sect. 4.2, the minimax in (5.10) is achieved when the density takes the form

$$P_\Lambda(l) = c\left[\sum_i \lambda_i q_i^2 \sigma_i^2 \exp\{-2\sigma_i l\}\right]^{1/2}, \quad \Lambda = \{\lambda_i\},$$

for the vector $\Lambda = \Lambda^*$, minimizing the function

$$G(\Lambda) = \left\{\int_0^\infty \left[\sum_i \lambda_i q_i^2 \sigma_i^2 \exp\{-2\sigma_i l\}\right]^{1/2} dl\right\}^2$$

under the condition that

$$\sum \lambda_i = 1, \quad \lambda_i \geq 0, \quad i = 0, \ldots, 11.$$

The value $G(\Lambda^*)$ then equals the minimax in (5.10). In Sect. 4.4 we mentioned that at $q_i \equiv 1$ the minimax measure Λ^* for a set of exponents is concentrated on the end points of the interval of values of σ, i.e.,

$$\Lambda^* = (\lambda^*, 0, \ldots, 0, 1 - \lambda^*)$$

and the density as in (5.9) is effectively optimal.

In this case only the values of $q_0, q_1, \ldots, q_{11}$ are strongly varied (these values are less remarkable than the remaining set of the values). Therefore, we calculated the quantity

$$G^*_{2,10} = \max_x \{G(\Lambda) | \lambda_2 = \lambda,\ \lambda_{10} = 1 - \lambda,\ 0 < \lambda < 1\} = 0.716$$

and the corresponding value $\lambda = \lambda^* = 0,30$. On the other hand, it appears that

$$F(p) = \max_i \int_0^\infty \frac{q_i^2 \sigma_i^2 \exp\{-2\sigma_i l\}}{p(l)} dl = 0.721$$

for the density (5.11). The inequality

$$G^*_{2,10} \leq G(\Lambda^*) \leq F(p)$$

fulfils the exact value of the minimax (5.10), consequently the density (5.11) is effectively optimal.

We calculated the ratios of the values of the spectrum for the particles escaping the ball to the corresponding values of the fission spectrum; these ratios varied from 0.4 to 1.4. In 14 s. of computer time we achieved an error of order 0.02-0.03. Notice that the direct simulation technique requires ~100 times as much computer time.

In Monte Carlo calculations one often uses the so-called run (or path) estimate for which the functions h depend on two successive points of the collision chain, i.e., $h = h(r_n, r_{n+1})$. For these estimates it is possible to construct weighted modifications by introducing a new chain $y_0, \ldots, y_n, \ldots$, where $y_n = (r_n, r_{n+1})$. The corresponding equations for the mean squared estimates take the form of (5.2) (Sect. 2.2).

5.3 Vector Algorithm Related to a Stratified Sampling with Respect to One Variable

The integral equation $\varphi = K\varphi + h$ can be written in the form of a system [5.5]:

$$\varphi(x) = \sum_{j=1}^{m} \int_{D_j^{(m)}} k(x,y)\varphi(y)dy + h(x), x \in D_j^{(m)}, \quad i = 1, ..., m. \tag{5.12}$$

This system could be reduced to a system as in (1.1) by a change of variables. In [5.5] we obtained some asymptotic estimations of the cost of the corresponding vector algorithm when the domains $D_i^{(m)}$ are uniformly diminished. If the dimension of the problem is larger than one, then there exists an optimal value of m. This means that when the method of finite sums is combined with the method of simple iterations, it is useful to choose a fixed value of m and then to diminish the error by averaging over all random realizations. Notice that in [5.5] we formaluted this approach more generally. Instead of the random point x_n, a random vector X_n with components $x_n^{(i)} \in D_i (i = 1, ..., n)$ is used and, in the weight factor, instead of $p(x_{n-1}, x_n)$, the corresponing marginal densities of $x_n^{(i)}$ are used (Sect. 5.10).

Let us now consider the case when $x = (z, \mu)$, where the variable z varies in the interval $(0, T)$. We suppose that all the functions in the equation are nonnegative. To construct (5.12), we divide the layer $X = \{(z, \mu) : 0 \le z \le T\}$ in $D_i^{(m)} (i = 1, ..., m)$ layers of equal depth T/m. A shift of variables reduces the layers to one layer $D^{(m)} = \{(z, \mu) : 0 \le z \le T/m\}$.

We then obtain a system

$$\varphi(x_i) = h(x_i) + \sum_{j=1}^{m} \int_{D^{(m)}} k(x_i, y_j)\varphi(y_j)dy_j, i = 1, ..., m, \tag{5.13}$$

where $y_j = (z_j', \mu'), z_j' = z' + (j-1)T/m, y = (z', \mu'), 0 \le z' \le T/m$, and x is defined similarly where the prime is omitted. Notice that $y_j \in D_j$, $y \in D^{(m)}$.

Let $p_m(x, y)$ be the corresponding vector estimate for solving (5.13). We now obtain an upper bound for the variances on the basis of the matrix-integral equation (5.2) and the Schwartz inequality

$$\mathrm{M}(\xi_{yi}^{(m)} \xi_{yj}^{(m)}) \le [\mathrm{M}\xi_{yi}^{(m)2} \mathrm{M}\xi_{yj}^{(m)2}]^{1/2}.$$

It follows that the matrix $\Psi(y)$ is bounded by $\Psi_y^{(m)} \Psi_y^{(m)\prime}$. Here $\Psi_y^{(m)}$ is a vector with the coordinates $[\mathrm{M}\xi_{yj}^{(m)2}]^{1/2}$ $(j = 1, ..., m)$. Simple arguments show that

$$\mathrm{M}\xi_{xi}^{(m)2} \le \chi(x_i) + \int_{D^{(m)}} \left[\sum_{j=1}^{m} k(x_i, y_j) \left[\mathrm{M}\xi_{y,j}^{(m)2} \right]^{1/2} \right]^2 p_m^{-1}(x, y)dy. \tag{5.14}$$

Let us now consider the equation

$$g_m(x) = \chi(x) + \int\limits_{D^{(m)}} \left[\sum_{j=1}^{m} k(x,y) g_m^{1/2}(y_j) \right]^2 p_m^{-1}(x_D, y) dy, \quad x \in X. \quad (5.15)$$

The right-hand side of this equation is obtained by "sewing" the right-hand sides of (5.24) with respect to i; the value of x_n is determined by the equality $z = z_D + (i-1)T/m (x \in D_i)$.

Assume that there exists a unique bounded solution to (5.15) which can be obtained by successive iterations starting from an arbitrary $g^{(0)} \in L_\infty(X)$. Then from Lemma 1.1 and (5.14) it follows that $\mathbf{M}\xi_{x_D,i}^{(m)2} \leq g_m(x)$, which can be rewritten as

$$\mathbf{M}\xi_x^{(m)2} \leq g_m(x), x \in X, \quad (5.16)$$

where $\xi_x^{(m)2}$ is obtained by "sewing" the quantities $\xi_{xi}^{(m)2}$. The assumption made above follows from the results of Sect. 3.2 under the condition of the limit value of $\mathbf{M}\xi_x^{(m)}2$. In Theorem 5.2, the following representations for the kernel $k(x,y)$ and the transition density $p(x,y)$ are used:

$$k(x,y) = k^{(1)}(x,\mu')k^2((z,\mu'),z'), \quad \int k^{(2)} dz' = 1.$$

In addition, the increase of m is carried out by introducing an additional subdivision, e.g., $m = 2^k$.

Theorem 5.2. Assume that the following conditions are satisfied:

1)
$$p_m(x,y) = p^{(1)}(\mu,\mu')p_m^{(2)}(\mu,z'),$$
$$p_m^{(2)}(\mu,z) = p^{(2)}(\mu,z) \left[\int\limits_0^{T/m} p^{(2)}(\mu,z) dz \right]^{-1},$$

where the density $p^{(2)}$ satisfies the following property (uniformity property): restriction of $p^{(2)}$ on an interval is invariant (to within the shift of the variable) with respect to the position of this interval.

2) For almost all (z,μ') the total number of points of discontinuity of $k^{(2)}((z,\mu'),z')$, $p^{(2)}(\mu',z')$, $g_m(y)$ $(m = 1,2,...,)$ as functions of z', is finite, and the density $p^{(2)}$ is positive and continuous at $z' = 0$.

3) $g_m(x) < c < \infty$ for some value of m, and

$$k^{(2)}((z,\mu'),z') < c(\mu') < \infty;$$

4) $$\int \frac{k^{(1)2}(x,\mu')}{p^{(1)}(\mu,\mu')} d\mu' \leq q < 1.$$

Then

$$\lim_{m\to\infty} \mathbf{M}\xi^{(m)2} = G(x) \le g(x),$$

where $g(x)$ is the unique in L_∞ solution to the equation

$$g(x) = \chi(x) + \int \frac{k^{(1)2}(x,\mu')}{p^{(1)}(\mu,\mu')} d\mu \left[\int_0^T k^{(2)}((z,\mu'),z') g^{1/2}(y) dz' \right]^2 . \qquad (5.17)$$

Proof. It is clear that under condition 1 the estimate $\xi_x^{(m)}$ is obtained by conditional averaging over new subdivisions. Therefore,

$$\mathbf{M}\xi_x^{(m)2} \downarrow G(x), \quad m \to \infty.$$

Equation (5.15) determines the mean square of a scalar estimate for the following method of simulation of z': first, an optimal choice of the layer j is carried out, next in $D_j^{(m)} z'$ is sampled according to $p_m (m = 1,2,...)$ (see Sect. 3.2, where it is proved that there exists a unique solution to (5.17) under condition 4). Therefore

$$g_m(x) \downarrow g^*(x) \le c.$$

The integral in (5.15) can be rewritten as follows (which follows from assumptions 2), 3):

$$\int \frac{k^{(1)2}(x,\mu')}{p^{(1)}(\mu,\mu')} d\mu' \times \left\{ \sum_{j=1}^{m} k^{(2)}((z,\mu'),z_j^*) g_m^{1/2}(y_j^*) \frac{T}{m} \left[1 + O\left(\frac{T}{m}\right)\right] \right\} . \qquad (5.18)$$

By the Egorov theorem [5.9] the convergence $g_m(x) \downarrow g^*(x)$ is almost uniform; therefore, (5.18) converges to the integral in (5.17) where g is replaced with g^*. Thus, it is possible to take the limit in (5.16); the theorem is proved. □

Note that the exponential and uniform densities can be regarded as realistic variants of $p^{(2)}$. However for large values of However, for large values of m (small depth of the layer) the densities are essentially equivalent to uniform densities.

Note also that (5.17) corresponds to a scalar algorithm with optimal simulation of the run of the particle. It follows that if the scattering is simulated by the importance sampling technique (Sect. 2.1), then $g(x) = (\mathbf{M}\xi_x)^2$ (Sect. 3.2); thus the variance of the vector estimate tends to zero as $m \to \infty$.

As an example, let us consider the one-velocity transfer equation for a plane–parallel layer $0 \le z \le T$ with the scattering indicatrix $w(\mu,\mu')$, the mean free length $1/\sigma = 1$, the survival probability q (μ is the cosine of the angle between the axis z and the direction of the particle). For this equation (Sect. 1.3)

$$k(z,\mu;z',\mu') = w(\mu,\mu') \exp\{-(z'-z)/\mu\}/\mu|^{-1} \quad ,$$

where $(z'-z)\mu > 0$ and $k(z,\mu;z',\mu') = 0$ otherwise. Here $-1 \le \mu \le 1, 0 \le z \le 1$. If we take $h(z,\mu) = \exp\{-|T-z|/\mu\}$ for $\mu > 0$ and $h(z,\mu) \equiv 0$ otherwise, then $\varphi(z,\mu)$ can be interpreted as the probability that a particle escapes going through the boundary $z = T$ in the case of a unique point source located at the point (z,μ).

We divide the layer $0 \le z \le H$ into m equal sublayers of depth T/m. The vector algorithm for solving the system consists of the following: the collision point is sampled in the layer and the unbiased estimate φ is constructed on the basis of the corresponding matrix weight.

Heuristic arguments [5.5, 10] give the following algorithm (A) for simulating the run of the particle (i.e., the choice of a new collision); the particle moves along its direction until the boundary of the elementary layer, then the particle jumps to the opposite boundary and continues to move in the same direction. The length of the run has the exponential distribution on the interval $(0, T/|m\mu|)$. The corresponding density $p_n(z)$ takes the form

$$p_n(z) = \begin{cases} \dfrac{\exp\{-(z-z_{n-1})/\mu_{n-1}\}}{a_{n-1}\mu_{n-1}}, & z_{n-1} \le z \le T/m, \\[2ex] \dfrac{\exp\{-(-z+T/m-z_{n-1})/\mu_{n-1}\}}{a_{n-1}\mu_{n-1}}, & 0 \le z \le z_{n-1}, \end{cases}$$

where $a_{n-1} = 1 - \exp\{-T/(m\mu_{n-1})\}$. The spatial weight factor in this case takes the form

$$q_{ji} = \begin{cases} a_{n-1}\left[\exp\left\{-\dfrac{T}{m\mu_{n-1}}\right\}\right]^{(i-j)}, & i > j \cap z_n \le z_{n-1}, \\[2ex] a_{n-1}\left[\exp\left\{-\dfrac{T}{m\mu_{n-1}}\right\}\right]^{(i-j-1)}, & i > j \cap z_n \le z_{n-1}, \\[2ex] 0, & i < j \cup (j = i \cap z_n \le z_{n-1}). \end{cases}$$

Analogously, the density $p_n(z)$ and the weight factor are constructed in the case $\mu_{n-1} < 0$. It is possible to use algorithm (B), which is simpler for theoretical investigations: the point z' is sampled from the density $c|\mu|^{-1}\exp\{-z/|\mu|\}$ on the interval $(0, T/m)$. The scattering is simulated according to $w(\mu,\mu')$ for both algorithms.

Lemma 5.3. For the algorithm (B) there exists m_0 such that $g_m(x) < c < \infty$ for $m > m_0$.

Proof. As mentioned above, $g_m(x)$ is equal to the mean square of the scalar estimate which is obtained as a result of the optimal choice of the layer when simulating z'. It is obvious that the first step of such optimization can be constructed on the basis of the restriction of the density on the layers which are intersected by the particle. As a result, we obtain algorithm B_0 with weights; the ratio of these weights to the weights of the direct simulation estimate without escape [5.7, p.176] tends to 1 as $m \to \infty$. The variance of the direct simulation technique is finite; therefore, the variance of algorithm B_0 will also be finite beginning with some value of m. Thus, the quantity $g_m(x)$ which corresponds to the total optimization of the choice of the layer is finite. □

Lemma 5.4. Algorithm (A) is optimal according to the criterion (5.6) asymptotically, as $T \to \infty$.

Proof. Assume that $\mu > 0$. Then

$$\sum_{j=1}^{m} k_{1j}(x,y) \geq \sum_{j=1}^{m} k_{ij}(x,y), \quad i = 1, ..., m.$$

Consequently, to reach the minimax, it is sufficient to minimize the first integral in (1.6). Then, using the importance sampling principle, we conclude that the optimal density has the form

$$p^*(x,y) = c\sum_{j=1}^{m} k_{ij}(x,y) = cp_A(x,y)\left[1 + O\left(\exp\left\{-\frac{T}{|\mu|}\right\}\right)\right],$$

where $p_A(x,y)$ is the transition density of algorithm A. □

We now study the application of Theorem 5.2 to algorithm (B). Conditions 1), 2) are obviously satisfied, and assumption 3) is also true due to Lemma 5.3. Thus, for sufficiently large values of m the variance of algorithm (B) is, in effect, not larger than that of the scalar algorithm with an optimal simulation of the free length. If, in addition, the scattering is simulated on the basis of the importance sampling technique, then the variance is close to zero.

It is clear that asymptotically (as $m \to \infty$) algorithm (A) is equivalent to algorithm (B), but for $m = 1$ algorithm (A) is essentially better. It is better also from the computational point of view, since the calculations of the weights are simple. So we conclude, using Lemma 5.4, that the algorithm (A) is preferable.

In [5.5] we solved a problem of radiative transfer through a plane-parallel layer of material with isotropic scattering and $T = 20$, $g = 0.8$. The particles started at the plane $z = 0$ in the direction of the z-axis [i.e., $x = (0,1)$]. Monte Carlo calculations of the transmission probability $P = \mathbf{M}\xi_m$ were carried out for various values of the parameter m, and the variance $\sigma_m^2 = \mathbf{D}\xi_m$ was calculated simultaneously. Repeated calculations have given $\mathbf{P} \approx 0.91 \cdot 10^{-6}$. It is well-known that a strong positive correlation between the statistical estimates of the expectation and the variance exists . Therefore, we improve the estimate of σ_m on the basis of multiplication by $0.91 \cdot 10^{-6}$ divided by the estimate of $\mathbf{M}\xi_m$. The results are given in Table 5.1, where t_m is the computer time of simulation of one trajectory (in seconds); the optimal value of m is equal to 5.

Table 5.1

m	t_m	$\sigma_m \cdot 10$	$t_m \cdot \sigma_m^2 \cdot 10$
1	47	363	620
2	58	341	675
4	88	86	65
5	107	77	63
6	122	74	67
7	145	69	69
10	247	54	72
20	749	46	157
25	1112	44	211

The algorithm described in this section is also applicable in the case where the material is nonhomogeneous in z-coordinate and has an arbitrary scattering indicatrix which changes the energy. This algorithm has an important advantage compared to other Monte Carlo methods when it is necessary to calculate the z-distribution of the flux of the particle.

For $m = 20$ we calculate in addition to the probability P the probability P_1 that a particle will transmit a layer $0 \leq z \leq 10$:

$$P \approx P^{(1)} = 0.9116 \cdot 10^{-6}, \quad P_1 \approx P_1^{(1)} = 0.1094 \cdot 10^{-2}.$$

The mean squared errors σ of these estimates were 16% and 8%, respectively. However, the difference between the quantity

$$P^{(1)}/P_1^{(1)} = 0,833 \cdot 10^{-3}$$

and $P/P_1 = 0.825 \cdot 10^{-3}$ calculated on the basis of the asymptotic solution (Sect. 3.3) was 1%.

An analogous algorithm can be constructed for calculating the patricle's transfer for a system with a spherical symmetry. It would be advantagous to evaluate the flux of the particle at the center of the sphere.

Note that the approach described cannot be applied in solving the general transfer equation (Sect. 1.3), since the kernel includes a δ-function which relates the collision point to the direction of the previous run. However, the equation $f = Kf + \Psi$ can be rewritten as

$$f + K^2 f + K\Psi + \Psi$$

with the iterated kernel

$$k_1(x', x) = \int_X k(x', x'')k(x'', x)dx'',$$

which is regular. The integrals which then appear in the expressions for the weights can be calculated by the Monte Carlo method.

In conclusion, we note that the scale of integration in (3.6) is approximately equal to the correlation scale of the quantities $\xi_x^{(m)}$; therefore $\mathbf{M}\xi_x^{(m)2}$ and $g(x)$ are probably very close (in the limit). For this example we obtained

$$\left[\frac{g(x)-\varphi^2(x)}{\mathbf{M}\xi_x^{(m)2}-\varphi^2(x)}\right]^{1/2} \approx 1.23, \quad m \geq 20.$$

This implies that the vector algorithm is, in fact, not better than the scalar algorithm with an optimal simulation of the free length, if the pointwise estimate for $\varphi(x)$ is constructed for one variant of $h(x)$. However, the vector algorithm gives estimates for $\varphi_i(x)(i = 1, ..., m)$ which have strong positive correlations; thus, it would be effective in the problems where it is necessary to evaluate the function $\varphi(x)$ in many points or when a set of equations with different functions $h(x)$ must be solved.

5.4 Accuracy of the Monte Carlo Method for Solving the Vector Transfer Equation

We now consider vector integral equations of radiative transfer, considering polarization (Sect. 1.3):

$$\varphi_i(x) = \int \sum_{j=1}^{4} k_{ij}(x,y)\varphi_j(y)dy + h_i(x), \quad i = 1, ..., 4. \tag{5.19}$$

We use the following notation: $H(x)$ is a column vector function, $H(x) = (h_1(x), ..., h_4(x))'$, and $K(x,y)$ is the matrix of kernels of the system (5.19). Here $x = (r, \omega)$ is a phase point of the space R^3, $\omega \in \Omega$ – direction unit vector, and $y = (r', \omega')$. Compared with Sect. 1.3, (5.19) is written in an adjoint form, therefore we can write

$$k(x,y) = K(x, (r_y, \omega))P'(\omega, \omega_y),$$

where k is a substochastic density of transition from r to r' along ω (Sect. 1.3), and

$$P'(\omega, \omega_y) = L(-i_2)R(\mu)L(\pi - i_2)/(2\pi), \quad \mu = (\omega, \omega_y),$$

$$R(\mu) = \begin{bmatrix} r_{11} & r_{12} & 0 & 0 \\ r_{21} & r_{22} & 0 & 0 \\ 0 & 0 & r_{33} & r_{34} \\ 0 & 0 & -r_{43} & r_{44} \end{bmatrix}, \quad L(i) = \begin{bmatrix} 1 & 0 & 0 & 0 \\ 0 & \cos 2i & \sin 2i & 0 \\ 0 & -\sin 2i & \cos 2i & 0 \\ 0 & 0 & 0 & 1 \end{bmatrix},$$

where

$$r_{ij} = r_{ij}(\mu), \quad r_{11} \geq 0, \quad \int_{-1}^{1} r_{11}(\mu)d\mu = 1.$$

It is assumed that the medium is isotropic and that P is independent of r.

To solve (5.19) by the Monte Carlo method, one constructs a random estimate ξ_x such that

$$\mathbf{M}\xi_x = \Phi(x),$$

where the transition density has the form

$$p(x,y) = p_1(x,(r_y,\omega))p_2(\mu)/(2\pi).$$

Equation (5.2) for the covariance matrix $\Psi(x) = \mathbf{M}(\xi_x, \xi'_x)$ is considered in $E(R\times\Omega)$, the space of functions continuous on $R^3\times\Omega$ matrix-valued functions with the norm

$$\|\Psi\| = \sup_{i,j,x} \|\Psi_{ij}(x)\|.$$

Note that (5.2) holds if $\rho(K_p) < 1$. The last condition is weaker than that of Lemma 5.1. Indeed, if $\rho(K_p) < 1$, then $\mathbf{M}\xi_{x1}^2 < \infty$. In addition, from the properties of the Stokes vector functions (Sect. 1.3), we have

$$\xi_{x1} \geq 0, \quad |\xi_{xi}| < c\xi_{x1}, \quad i = 1,...,4.$$

Therefore, the iterated averaging in Lemma 5.4 is justified; hence (5.2) holds.

It is not difficult to see that the matrix-integral operator S, obtained from K_p by replacing $x \to \omega, y \to \omega_y, p \to p_2, K \to P'$, is compact if

$$r_{ij}(m\mu)[p_2(\mu)]^{-1/2} \in C[-1,1], \quad i,j = 1,...,4, \tag{5.20}$$

where $C[-1,1]$ is the space of functions continuous on the interval $[-1,1]$.

Lemma 5.5. The following inequality holds:

$$\rho(K_p) \leq q\rho(S_p),$$

where

$$q = \sup_{x,w} \int \frac{k^2(x,(x+l\omega,\omega))}{p_1(x,(x+l\omega,\omega))} dl.$$

Proof. Note that the integration is carried out with respect to the length of the transition from the point x in the direction ω. Consider now the first component of the nth iteration of the operator K_p and estimate consecutively the integrals taken with respect to the length of the first, second, third, etc., transitions. Making use of the properties of the Stokes vector, we find

$$\|K_p^n\|^{1/n} \leq c^{1/n}q\|S_p^n\|^{1/n}.$$

Now $n \to \infty$, completing the proof. □

Note that the value of the coefficient q is overestimated and the actual value can coincide with $\rho(L_p)$, where L_p is an integral operator with the kernel

$k^2 p_2/p_1$ (see below). If $q < 1/\rho(S_p)$, then estimates of the form $F'\xi_x/\pi$ have finite variance (Sect. 5.1). We now turn to evaluating the quantity $\rho(S_p)$.

Lemma 5.6. Assume that (5.20) is satisfied. Then the cone $T \subset E(\Omega)$ of functions continuous on Ω whose values are nonnegative definite matrices is invariant under the transformation S_p, and $\rho(S_p) = \lambda_0$, where $\lambda_0 > 0$, is the maximal positive eigenvalue of the operator S_p.

Proof. Note that $K\Psi K'$ transforms a nonnegative definite matrix Ψ to a nonnegative definite matrix, and (5.20) ensures continuity of the matrix kernel of (5.2). The operator S_p is compact and $L(T) = E(\Omega)$, where $L(T)$ is the closure of a linear span; hence $\rho(S_p)$ in the positive eigenvalue. On the other hand, the compactness yields $\rho(S_p) = \lambda^*$, where λ^* is the maximal absolute value of the eigenvalues of S_p. This proves Lemma 5.6. □

We will now seek the eigenmatrix $\Psi^{(0)}(\omega)$ of the operator S_p in the form of a diagonal matrix with nonnegative diagonal elements $(1, a_1, a_1.a_2)$. It is easy to see that $S_p\Psi^{(0)}$ is diagonal. Therefore, we obtain

$$\begin{aligned} &c_{11} + c_{12}a_1 = \lambda_0 \\ &c_{12} + (c_{22} + c_{33})a_1 - c_{34}a_2 = 2\lambda_0 a_1 \\ &c_{43}a_1 + c_{44}a_2 = \lambda_0 a_2, \end{aligned} \tag{5.21}$$

where

$$c_{ij} = \int [r_{ij}(\mu) r_{ij}(\mu)/p_2(\mu)]d\mu.$$

Lemma 5.7. If there exists a solution to (5.21) with positive components λ_0, a_1 and a_2, then λ_0 is the maximal positive eigenvalue of S_p.

Proof. By the hypothesis of Lemma 5.7, λ_0 is a positive eigenvalue of S_p corresponding to a diagonal eigenmatrix Ψ_0 with positive elements. As Ψ_0 is an inner element of the cone T, the quantity λ_0 is the maximum among the positive eigenvalues of K_p, which proves the statement (Sect. 5.9). □

Theorem 5.3. If the conditions of Lemmas 5.5, 6, 7 hold, then $\rho(S_p) = \lambda_0$, where λ_0 is the positive solution to (5.21). □

The relations obtained allow us to solve numerically the problem of minimizing the quantity $\rho(S_p)$ as a functional of p_2 chosen from the class

$$p_2(\mu) = c\left[\sum_{i,j=1}^{4} \lambda_{ij}|r_{ij}(\mu) r_{ji}(\mu)|\right]^{1/2}, \quad \lambda_{ij} \geq 0,$$

on which the minimax of the coefficients c_{ij} for S_p is realized. On finding an optimal value $\{\lambda^*_{ij}\}$, the minimax value can be used as an input value.

We consider now the case of Rayleigh scattering [5.3]. The Rayleigh scattering matrix is characterized by $r_{11} \equiv r_{22}, r_{21} \equiv r_{12}, r_{33} \equiv r_{44}, r_{34} \equiv 0$, with $r_{11}^2 = r_{12}^2 + r_{33}^2$. Equation (5.21) can be rewritten as

$$c_{11} + c_{12}a_1 = \lambda_0, \quad c_{12} + (c_{11} + c_{33})a_1 = 2\lambda_0 a_1, \lambda_1 = c_{33}. \tag{5.22}$$

If we take $p_2 \equiv r_{11}$, then $c_{11} = 1$; hence $a_1 = 1/2, \lambda_0 = 1 + c_{12}/2$. For Reyleigh scattering we have

$$r_{11}(\mu) = 3(1 + \mu^2)/8, \quad r_{12}(\mu) = 3(1 - \mu^2)/8, \quad r_{33}(\mu) = 3\mu/4,$$

and we find for $p_2 = r_{11}$

$$\lambda_0 = 1 + (3\pi - 8)/8 = 1.178, \quad \lambda_1 = 0.643 \quad .$$

The minimization procedure considered above for λ_0 reduces here to minimizing $\lambda_0(a, b)$ of the solution of (5.22), where

$$p_2(\mu) = c(1 + a\mu^2 + b\mu^4)^1/2.$$

In computations, $\lambda_0(0.7, 1.0) = 1.171$ proved to be the minimal value, and $1.71 \leq \lambda_0 \leq 1.172$ for all points from the interval

$$a(t) = 1.7t, \quad b(t) = 1.7(1 - t), \quad 0 \leq t \leq 1.$$

It is interesting that the minimax of the coefficients c is realized at $a = 2, b = 1$, i.e., for $p_2 \equiv r_{11}$. This explains why the variant $p_2 \equiv r_{11}$ is so effective.

Note that the spectral radius of the operator K_p can be evaluated approximately by the Monte Carlo calculations of the iterations of K_p [5.7]. More precisely, consider linear functionals of the form

$$a_n = \int F'(x)[K_p^n \Psi^{(0)}](x)F(x)dx,$$

where F is a vector function.

It is natural to choose $\Psi^{(0)}$ as a diagonal matrix with elements $(1, a_1, a_1, a_2)$, i.e., the eigenmatrix for an infinite homogeneous medium. Then $\Psi_{11} \leq \Psi_{22} + \Psi_{33} + \Psi_{44}$ holds for all iterations, which implies that this property is also true for the eigenmatrix obtained by the iterative process. Let us describe the Monte Carlo algorithm for evaluating a_n. Denote by $a_n^{(k)}$ the quantity obtained from a_n by replacing $\Psi^{(0)}$ with matrix $H_k H_k'$, where H_k is the corresponding basis vector, e.g., $H_1 = (1, 0, 0, 0)$. Therefore, we can write [5.1]

$$a_n^{(k)} = \mathbf{M}\left[H_k'(x_n)Q_n' \frac{F(x_0)}{\pi(x_0)}\right]^2,$$

where $\pi(x_0)$ is the distribution density of the initial state of the Markov chain, and Q_n are constructed as described in [5.1]. Hence, $a_n^{(k)}$ is the mean squared value of the k-th component of the weight Stokes vector obtained on the n-th state of the Markov chain, as described in [5.3], and

$$a_n = a_n^{(1)} + a_1 a_n^{(2)} + a_1 a_n^{(3)} + a_2 a_n^{(4)}.$$

For compact operators $K_p, a_{n+1}/a_n \rightarrow \rho(K_p)$ as $n \rightarrow \infty$. Numerical experiments show that for $p_2 \equiv r_{11}$ this technique gives approximately the same value of $\rho(S_p)$ as that obtained for λ_0 from (5.21).

For a plane layer of medium with optical depth equal to 2 and the Rayleigh scattering without absorption, the iterative technique gives

$$\rho(K_p) \approx 0.93, \quad q = \rho(K_p)/\lambda \approx 0.79, \quad \rho(L_p) \approx 0.80 \quad ,$$

where $p_1 \equiv k, p_2 \equiv r_{11}$, and L_p is a scalar integral transfer operator with the indicatrix $r_{11}(\mu)$. The agreement $\rho(K_p) \approx \rho(S_p)\rho(L_p)$ can probably be explained by the following arguments. The maximal eigenvalue of the operator K_p can be written in the form

$$\lambda_0^{(p)} = (\Psi^*, K_p\Psi_0)/(\Psi^*, \Psi_0), \tag{5.23}$$

where Ψ^* is the eigenfunction of K_p^* corresponding to $\lambda_0^{(p)}$ and Ψ_0 as defined above. A simple analysis shows that by taking $\Psi^* = \Psi_0^* \varphi^*(x)$, where $\varphi^* = \lambda L_p^* \varphi^*$ and $\lambda = \rho(L_p)$, we obtain from (5.23)

$$\lambda_0^{(p)} = \lambda\lambda_0.$$

Note that the perturbation theory was employed. As a rule, perturbation techniques are quite effective in transfer theory.

5.5 Vector Estimates for Triangular Matrix Kernel

Consider a system of the form

$$\chi_i = h_i + \sum_{j=1}^{i} K_{ij}\chi_i, \quad i = 1, ..., m, \tag{5.24}$$

where K_{ij} is an integral operator with the kernel $k_{ij}(x,y)$.

Theorem 5.4. If K_{ij} are bounded operators, then the inequality

$$\rho(\mathbf{K}) \leq \max_i\{\rho(K_{ii})\} = \rho_0 \tag{5.25}$$

holds. Furthermore, if $k_{ij}(x,y) \geq 0$, then $\rho(\mathbf{K}) = \rho_0$.

Proof. Let us transform (5.24) as follows: replace $\mathbf{K} \rightarrow \lambda\mathbf{K}$, where $\lambda > 0$, i.e., consider the system

$$\chi_i = h_i + \lambda \sum_{j=1}^{i} K_{ij}\chi_i, \quad i = 1, ..., m. \tag{5.26}$$

Consider also the system

$$\chi_i^{(t)} = t^{i-1}h_i + \lambda \sum_{j=1}^{i} t^{i-j} K_{ij}\chi_i^{(t)}, \quad i = 1, ..., m, \tag{5.27}$$

or

$$\chi^{(t)} = t^{i-1}H + \lambda \mathbf{K}^{(t)}\chi^{(t)}$$

for an arbitrary $t > 0$. It is easy to show by induction that the solutions of (5.26, 27) are related as follows: $\chi_i^{(t)} = t^{t-1}\chi_i$, and the i-th component of the Neumann series for (5.27) can be written as

$$\sum_{n=0}^{\infty} \lambda^n (\mathbf{K}^{(t)n} H)_i = \sum_{n=0}^{\infty} \lambda^n t^{i-1} (\mathbf{K}^n H)_i.$$

The Neumann series for (5.27) thus converges or diverges simultaneously for all values of $t > 0$. Considering the norms of $\mathbf{K}^{(t)n}$ and using the boundedness of the operators K_{ij}, we obtain

$$\lambda^n \|\mathbf{K}^{(t)n}\| \to \lambda^n \sup_i \{\|K_{ii}^n\|\} \quad \text{as } t \to 0. \tag{5.28}$$

On the other hand, if $\lambda < \rho_0^{-1}$, then

$$\lambda^n \sup_i \{\|K_{ii}^n\|\} \to 0, \quad \text{as} \quad n \to 0.$$

Hence, the Neumann series for (5.27) converges for all $t > 0$ (i.e., for (5.26) as well) if $0 < \lambda < \rho_0$. Therefore, $\rho(\mathbf{K}) < \rho_0$ [5.11]. If $k_{ij}(x, y) \geq 0$, then $\sup\{\|K_{ii}^n\|\} < \|\mathbf{K}^n\|$ and $\rho_0 \geq \rho(\mathbf{K})$. □

Note that the first statement of Theorem 5.4 holds for arbitrary linear operators. Denote by $K_{p,ij}$ an integral operator with kernel $k_{ij}^2(x, y)/p(x, y)$.

Theorem 5.5. If $K_{p,ij}$ are bounded operators, then (5.24) satisfies the inequality

$$\rho(\mathbf{K}_p) \leq \max_i \{\rho(K_{p,ij})\} = \rho_0^{(p)}.$$

If, in addition, $k_{ij}(x, y) \geq 0$, then $\rho(\mathbf{K}_p) = \rho_0^{(p)}$.

Proof. Repeating the proof of Theorem 5.4 and analysing the convergence of the Neumann series for (5.2) corresponding to (5.27), we find that it is sufficient to prove the following relation:

$$\lambda^{2n} \|\mathbf{K}_p^{(t)n}\| \to \lambda^{2n} \sup_i \{\|K_{p,ii}^n\|\}, \quad t \to 0. \tag{5.29}$$

Expressions (5.2, 3) show that (5.29) holds if operators with kernels of the form $k_{ij}(x,y)k_{rs}(x,y)/p(x,y)$ are bounded, which follows from the obvious inequality

$$\int \frac{k_{ij}(x,y)k_{rs}(x,y)}{p(x,y)}dy \leq \|K_{p,ij}\|^{1/2}\|K_{p,rs}\|^{1/2}. \qquad \square$$

It follows from Theorem 5.5 that the variances $\{\mathbf{D}\xi_{j,x}\}$ of the estimates of the solution components for (5.24) are finite if, for example, $\|K_{p,ij}\| < 1, i = 1,2,...,m$. This gives a simple condition of variance finiteness for the vector algorithm for the multi-group transfer equations proposed in [5.2]. Here, the transition into another group is treated as absorption. Therefore, the equality

$$\max_i \|K_{p,ii}\| = \|K_{p,11}\|$$

seems to be true when the spectrum of the problem is not too stiff, i.e., the maximum of the norms of operators $K_{p,ii}$ is realized on the lowest energy group [note that usual group indexing is inverse to that of (5.24)].

Note also that a system of integral equations related to the polyharmonic equation [5.6] also has a triangular form, and the diagonal elements are integral operators related to the Laplace operator. Therefore, we can conclude from Theorem 5.5 that the walk on spheres algorithm for the polyharmonic equation has a finite variance, since it has a finite variance for the Laplace equation.

5.6 Vector Estimates for the Resolvent Iterations

It is easy to verify that

$$\chi_i(x) = \prod_{k=1}^{i} t_k(x) + \sum_{j=1}^{i}\left[\prod_{k=j+1}^{i} t_k(x)\right]\lambda_j \int k_j(x,y)\chi_j(y)dy \tag{5.30}$$

is satisfied by functions $\chi_i (i = 1,...,m)$ which are related as follows:

$$\chi_j = \lambda_i K_i \chi_i + t_i \chi_{i-1}, \quad \chi_0 \equiv 1, \quad i = 1,...,m \tag{5.31}$$

where K_i is an integral operator. Rewrite (5.30) for $m = 3$ in the form

$$\chi_1(x) = t_1(x) + \lambda_1 \int k_1(x,y)\chi_1(y)dy,$$

$$\begin{aligned}\chi_2(x) =& t_1 t_2|_x + t_2(x)\lambda_1 \int k_1(x,y)\chi_1(y)dy \\ &+ \lambda_2 \int k_2(x,y)\chi_2(y)dy,\end{aligned}$$

$$\chi_3(x) = t_1t_2t_3|_x + t_2t_3|_x\lambda_1 \int k_1(x,y)\chi_1(y)dy + t_3(x)\lambda_2 \int k_2(x,y)\chi_2(y)dy + \lambda_3 \int k_3(x,y)\chi_3(y)dy.$$

The transition from the i-th to the $(i+1)$-th equation in (5.30) corresponds to (5.31), i.e., the right-hand side of the $(i-1)$-th equation is multiplied by t_1 and added to $\lambda_i K_i \chi_i$. Theorems 5.4, 5 are directly extended to (5.30).

In the next section we also consider a system of the form

$$\chi_i = K_i\chi_i + R_i\chi_{i-1}, \quad i = 1, ..., m \tag{5.32}$$

where R_i is an integral operator. Note that it is useless to consider (5.31) as a special case of (5.30), since we need to use an integral operator with a "delta-kernel", which invalidates the weight technique.

We put $k_i(x,y) \equiv k(x,y)$, $t_1(x) \equiv c_0 h(x)$, $\lambda_i \equiv \lambda$, $t_{i+1} \equiv t_0 (i = 1, 2, ..., m)$ and rewrite (5.30) as

$$\chi_i(x) = t_0^{i-1} c_0 h(x) + \sum_{j=1}^{i} t_0^{i-j} \lambda \int k_j(x,y)\chi_j(y)dy. \tag{5.33}$$

It will be shown that under an appropriate choice of parameters c_0, t_0 and λ the approach to the solution of (5.33) is equivalent to the realization of a special iterative process for solving integral equations of the first and second kind.

Note that (5.33) relates to

$$\chi_1 = \lambda K \chi_1 + c_0 h, \quad \chi_i = \lambda K \chi_i + t_0 \chi_{i-1}, \quad i = 2, ..., m,$$

i.e., the function $\sum_{i=1}^{m} \chi_i$ is a partial sum of the Neumann series for the equation

$$\varphi = t_0[I - \lambda K]^{-1}\varphi + c_0[I - \lambda K]^{-1}h. \tag{5.34}$$

Equation (5.34) is considered in L_∞, and the operator K is assumed to be bounded. If the value λ^{-1} does not belong to the spectrum $\sigma(K)$ of the operator K, then the following relation holds [5.11]

$$\sigma(t_0[I - \lambda K]^{-1}) = t_0[I - \lambda\sigma(K)]^{-1}.$$

Setting $t_0 = 1 - \lambda$ and $c_0 = \lambda$ in (5.34), we obtain $\varphi = K\varphi + h$; hence the Neumann series for (5.34) converges if

$$|\sigma(K)| \le 1, \quad |\sigma(K) - 1| \ge \varepsilon, \quad 0 < \lambda < 1.$$

Note that the last conditions could be extended. For example, if $\sigma(K) \geq\geq \rho > 1$, then it is possible to take $\lambda < 0$. In general, the values $\lambda \notin \sigma(k)$ must be chosen so that the spectral radii of the operator λK_1 with the kernel $\lambda|k(x,y)|$ and those of the operator K_p with the kernel $\lambda^2 k^2(x,y)/p(x,y)$ are less than 1. From this condition it follows that $\mathbf{M}\xi_{i,x} = \chi_i(x)$ and $\mathbf{D}\xi_{i,x} < \infty$, where $\xi_{i,x}$ is the i-th component of the vector estimate for the solution of (5.34). Therefore, the variance of the sum of quantities $\xi_{i,x}$ which estimates the partial sum of the Neumann series for (5.34) is also finite.

If we now put $t_0 = 1, c_0 = -\lambda$, (5.34) yields $K\varphi = h$. The technique described above for the integral equation of the second kind is applicable here if $|\lambda\rho(K_1)| < 1$ and $|1 - \lambda\sigma(K)| \geq r > 1$, for example, when $\sigma(K) \geq \varepsilon > 0$. An analogous estimate can be constructed if $\sigma(K)$ alternates in sign, taking $\lambda_k = (-1)^k\lambda(-1)^{1/2}, t_1 \equiv \lambda^2 h$ and adding the even values of $\xi_{i,x}$.

Theorem 5.6. Assume that $t_0 = 1 - \lambda > 0, c_0 = \lambda$, $0 < q < (1 - |\lambda|\|K_p\|^{1/2})$ $(1-\lambda)^{-1}$. Then

$$\mathbf{D}\xi_{ix} \leq c\|h\|^2(1 - \|K_p^{(q)}\|)^{-3/2}\lambda^2 q^{-2i}, \quad i = 1, 2, ..., \tag{5.35}$$

$$\|K_p^{(q)}\| \leq \lambda^2[1 - q(1-\lambda)]^{-2}\|K_p\| < 1.$$

Proof. Consider an infinite system as in (5.33) and replace $1 - \lambda$ by $q(1 - \lambda)$. It is easy to verify that

$$\|K_p^{(q)^2}\| \leq \frac{\lambda^2\|K_1\|^2}{[1 - q(1-\lambda)]^2} \leq \|K_p^{(q)}\| \leq \frac{\lambda^2\|K_p\|}{[1 - q(1-\lambda)]^2} < 1.$$

In addition,

$$\|\mu\| = \|H\chi' + \chi H' - HH'\| \leq 3\frac{\lambda^2\|h\|^2}{1 - \|K_1^{(q)}\|} \leq 3\frac{\lambda^2\|h\|^2}{(1 - \|K_1^{(q)}\|)^{1/2}}.$$

To complete the proof, it is sufficient to note that $\|\Psi\| \leq \|\mu\|(1 - \|K_p\|)^{-1}$, and the replacement $1 - \lambda \to q(1-\lambda)$ results in multiplying the estimate of $\xi_{x,i}$ by q^i. □

Consider now the cost $S(\varepsilon)$ of the method described, ε being the margin of error. The number m is defined by

$$m = c_1 \ln \varepsilon [\ln R(\lambda)]^{-1},$$

where

$$R(\lambda) = \rho(t_0[I - \lambda K]^{-1}).$$

Realization of the estimates $\xi_{i,x} (i = 1, 2, ..., m)$ needs $c_2 m$ operations, since $\xi_{i,x}$ can be represented as the product of $t_0^{i-1}\xi_{x,x'}$ and $K(x,y)$, a triangular matrix with identical nonzero elements. Assume that

$$\mathbf{D}(\zeta_{m,x}) \leq g(m),$$

where

$$\zeta_{m,x} = \sum_{i=1}^{m} \xi_{i,x} \quad .$$

Then $S(\varepsilon) \le \varepsilon^{-2} g(m) c_2 m$. The functions $S_m = \sum_{i=1}^{m} \chi_i$ satisfy

$$S_m = \lambda K S_m + t_0 S_{m-1} + c_0 h, \quad S_0 \equiv 0$$

and $\zeta_{m,x}$ are the components of vector estimates for the corresponding system of integral equations which is obtained from (5.33) by the following replacement:

$$t_0^{i-1} c_0 h(x) \to \sum_{k=0}^{i-1} t_0^k c_0 h(x).$$

This gives for $\mathbf{D}(\zeta_{m,x})$ an estimate of (5.35) (without the factor λ^2), if $Q = \|K_p\|^{1/2} > 1$ and $\lambda > 0$. Setting

$$t_0 = 1 - \lambda > 0, \quad c_0 = \lambda, \quad m \approx -(a|\lambda|)^{-1} \ln \varepsilon,$$

$$q \approx 1 - r\|\lambda\|(Q - \mathrm{sign}\lambda), \quad r > 1 \quad ,$$

we obtain

$$S(\varepsilon) < |\lambda^{-1} \ln \varepsilon| A(\varepsilon) \text{ as } \lambda \to 0 \quad ,$$

where

$$A(\varepsilon) = c_3 (r-1)^{-3} \varepsilon^{-2[1+(Q-sign\lambda)a^{-1}r]}.$$

Combining in $\zeta_{m,x}$ the terms with identical products of kernels $k(x_{n-1}, x_n)$, i.e., with identical scalar weights $Q_n^{(0)}$, we find

$$\zeta_{m,x} = \sum_{n=0}^{N} Q_n^{(0)} a_n^{(m)} h(x_n),$$

where $a_n^{(m)}$ can be evaluated in algorithm with $S(\varepsilon) \le A(\varepsilon)$. From (5.35), we obtain for $Q = 1, \lambda > 0, Q = 1 - \delta a \lambda, \delta > 0$ the estimate

$$A(\varepsilon) \le c_5 \delta^{-3} \varepsilon^{-2(1+\delta)}.$$

The inequality (5.35) also shows that it is possible to construct an unbiased estimate randomized with respect to m such that $S(\varepsilon) \le c\varepsilon^{-2}$. The random value of m is sampled according to the probability

$$p_m = (1-p)p^{m-1}, \quad p > q$$

and the quantities $\xi_{m,x}$ are added with weights $p^{-(m-1)}$.

5.7 Vector Representations of Bilinear Estimates

The Monte Carlo estimates for evaluating functionals of the form [5.8, 9]

$$(\varphi_1, \varphi_2^* t) = \int_X \varphi_1(x)\varphi_2^*(x)t(x)dx, \tag{5.36}$$

$$(\varphi_1, R^*\varphi_2^*) = \int_X \varphi_1(x)dx \int_X r(y,x)\varphi_2^*(y)dy, \tag{5.37}$$

are said to give "bilinear estimates" if the functions $\varphi_1(x)$ and $\varphi_2^*(x)$ satisfy the equations

$$\varphi_1(x) = \int_X k_1(x,y)\varphi_1(y)dy + h(x), \tag{5.38}$$

or

$$\varphi_1 = K_1\varphi_1 + h,$$

and

$$\varphi_2^*(x) = \int_X k_2(y,x)\varphi_2^*(y)dy + f(x), \tag{5.39}$$

or

$$\varphi_2 = K_2^*\varphi_2 + f,$$

where X is an n-dimensional Euclidean space, $t \in L_\infty(X)$, and (5.38, 39) are considered in $L_\infty(X)$ and $L_1(X)$, respectively. The operators K_1, K_2 and R are bounded in the L_∞-norm, and $\rho(K_1) < 1, \rho(K_2) < 1$. Functionals of the form (5.36, 37) are used in the Monte Carlo calculations of derivatives of the linear functionals with respect to the parameters of the problem considered (Sect. 1.8). Functional (5.36) can be represented as follows:

$$(\varphi_1, \varphi_2^* t) = ([I - K_2]^{-1} t [I - K_1]^{-1} h, f).$$

From this we obtain, using the results of Sect. 5.6,

$$(\varphi_1, \varphi_2^* t) = (\chi_2, f),$$

where χ_2 is determined from

$$\begin{aligned} \chi_1(x) &= h(x) + \int_X k_1(x,y)\chi_1(y)dy, \\ \chi_2(x) &= t(x)h(x) + t(x)\int_X k_1(x,y)\chi_1(y)dy + \int_X k_2(x,y)\chi_2(y)dy. \end{aligned} \tag{5.40}$$

The corresponding Monte Carlo algorithms can be constructed and investigated directly on the basis of Sections 1-3. Thus, the following statment holds. If $\rho(K_{p1}) < 1$ and $\rho(K_{p2}) < 1$, then $\mathbf{M}|\xi_{2x} = \chi_2(x), \mathbf{D}\xi_{2x} < \infty$. From this, it is not difficult to obtain an unbiased estimate for $(\varphi_1, \varphi_2^* t)$ with finite variance (Sect. 5.1) by introducing an appropriate randomization of x.

Functional (5.37) is represented in the form

$$(\varphi_1, R^* \varphi_2^*) = ([I - K_2]^{-1} R [I - K_1]^{-1} h, f) = (\chi_2, f),$$

and the correspondimg system of equations is written as

$$\chi_1 = h + K_1 \chi_1, \quad \chi_2 = R\chi_1 + K_2 \chi_2.$$

For variance finiteness, the operator R_p must be bounded.

It should be noted that the vector estimates for (5.36, 37) can be transformed to the weight bilinear estimates of Sect. 1.7. However, vector estimates are also convenient in computations. Note that the bilinear estimates of [5.12] were constructed only for the case $p(x,y) = k_2(x,y)$, which corresponds to the direct simulation of (5.38). In conclusion, we note that it is possible, in principle, to construct estimates for (5.36, 37) with zero variance, if all functional characteristics of the problem are nonnegative. This can be done by considering (5.40) to be a single integral equation where the sum of integrals in the seond equation is written as an integral with respect to a corresponding continuous discrete measure, and then using the importance sampling technique, i.e., by constructing $p(x,y)$ (and a randomization of x) in terms of $\chi(x)$ (Sect. 2.3).

In practice, the Monte Carlo estimates for derivatives of the linear functionals with respect to parameters of the problem involved are constructed by differentiating the corresponding linear estimates (Sect. 1.8). However, this differentiation must be justified. For example, differentiation of the known linear estimates of the transfer theory with respect to scattering and absorption coefficients is possible, and the variances of the resulting estimates are finite (Sect. 1.8) if the variances of random variables

$$\xi_1(\varepsilon) = \sum_{n=0}^{N} Q_n n (1+\varepsilon)^n \exp\left\{\varepsilon \sum_{k=1}^{n} l_k\right\},$$

$$\xi_2(\varepsilon) = \sum_{n=0}^{N} Q_n (1+\varepsilon)^n \left(\sum_{k=1}^{n} l_k\right) \exp\left\{\varepsilon \sum_{k=1}^{N} l_k\right\},$$

are also finite; l_k is the free path length after the $(k-1)$-th collision, and $\varepsilon > 0$ is a small arbitrary quantity. Usual methods [5.7] prove $\mathbf{D}\xi_i(\varepsilon) < \infty (i = 1,2)$ only for bounded medis, i.e. for $l_k < c < \infty$. However, Theorem 5.5 shows that the last condition is unnecessary for proving the finiteness of $\mathbf{D}\xi_1(\varepsilon), \mathbf{D}\xi_2(\varepsilon)$.

Indeed, it is sufficient to note that, for example, $\xi_2(\varepsilon)$ coincides with the estimate ξ_{2x} for

$$\chi_1(x) = 1 + \int_X k_\varepsilon(x,y) \chi_1(y) dy,$$

$$\chi_2(x) = \int\limits_X k_\varepsilon(x,y)\chi_1(y)dy + \int\limits_X |r_x - r_y| k_\varepsilon(x,y)\chi_2(y)dy.$$

Here r_x is the spatial component of a phase point x, and

$$k_\varepsilon(x,y) = (1-\varepsilon)\exp\{\varepsilon|r_x - r_y|\}k(x,y),$$

where $k(x,y)$ is the kernel of the transfer equation. Denote by K_p and $K_{p\varepsilon}$ the operators with kernels $k^2(x,y)/p(x,y)$ and $k_\varepsilon^2(x,y)/p(x,y)$, respectively. If $\rho(K_p) < 1$, then $\rho(K_{p\varepsilon}) < 1$; hence, by Theorem 5.5, $\mathbf{D}\xi_i(\varepsilon) < \infty, i = 1,2$ for sufficiently small ε. For example, direct simulations of the transfer process yield $\mathbf{D}\xi_i(\varepsilon) < \infty$ (for small ε) [5.7].

We now show that this approach gives the possibility of verifing the finiteness of the variances of the run estimates under general conditions, as mentioned in Sect. 2.2. Let us consider the system of equations

$$\chi_1(x) = 1 + \int\limits_X k(x,y)\chi_1(y)dy,$$

$$\chi_2(x) = \int\limits_X k(x,y)\chi_1(y)dy + \int\limits_X h(r_x, r_y)k(x,y)\chi_2(y)dy.$$

We assume that the elements of this system are nonnegative. Then

$$\xi_{2x} = \sum_{n=1}^{N} Q_n \left[\sum_{k=0}^{n} h(r_{k-1}, r_k)\right] \geq \sum_{n=1}^{N} Q_n h(r_{n-1}, r_n).$$

The last sum is, in fact, the weighted run estimate ζ_x introduced in Sect. 2.2. Consequently, by Theorem 5.5 $D\zeta_x < \infty$, if $\rho(K_p) < 1, \|K_p^{(h)}\| < \infty$, where $K_p^{(h)}$ is the operator with the kernel $h^2(r_x, r_y)k^2(x,y)/p(x,y)$. The condition $\|K_p^{(h)}\| < \infty$ is satisfied if the direct simulation is used and if $h(r_x, r_y) = |r_x - r_y|$.

In conclusion we note that there is a connection between the bilinear estimates and the class of unbiased estimates of the functionals of the type (φ_2^*, t). For a kernel $k_1(x,y)$ with $\rho(K_1) < 1$ we set

$$h(x) = 1 - \int\limits_X k_1(x,y)\, dy.$$

In addition, $\varphi_1 \equiv 1$, $(\varphi_1, \varphi_2^* t) = (\varphi_2^*, t)$. Consequently, there exists a set of unbiased bilinear estimates of functionals of type (φ_2^*, t) for which the finiteness of the variance is easily derived from the vector representation. It is not difficult to verify that the optimal estimate [according to (5.6)] in this case corresponds to $k_1 \equiv 0$, i.e., the standard collision estimate is optimal (Sect. 1.6).

5.8 Vector Algorithm for Evaluating the Effective Fission Coefficient

It is known [5.7] that the multiplication coefficient k_{eff} per one generation is equal to the maximal eigenvalue of the integral operator with a kernel $l(x, y)$ such that $l(x, y)dy$ is the mean number of particle collisions produced in a volume element dy by one collision at point x of the phase space in one generation [5.7], where the multiplication is treated as absorption. The coefficient k_{eff} is evaluated by the Monte Carlo method on the basis of

$$\frac{(L^m f, h)}{(L^{m-1} f, h)} \to k_{\text{eff}}, \quad \text{as} \quad m \to \infty.$$

Since the multiplication process may degenerate, one usually applies a modification, called the "method of generations with a constant number of points" [5.7]. This modification is known to be poorly adapted to the weight and correlated sampling techniques because of large weight fluctuations in the generation chain and the necessity of an additional intermixing of fissions in the construction of a new generation. Therefore, it is difficult to calculate small perturbations of the functionals studied and to decrease the variance of the estimates by the weight technique when the above generation modification is applied. In addition, calculations by this modification are difficult to parallelize.

Next we will consider a vector algorithm for calculating k_{eff} which is free of the disadvantages mentioned above.

Along with L, let us consider a standard integral operator of the transfer theory, K, with a kernel $k(x, y)$ such that $k(x, y)dy$ is the mean number of collisions in a volume element dy which occur immediately after collision at point x. It is clear that

$$L = cK^*[I - K^*]^{-1}, \quad c = \text{const}.$$

Consequently,

$$(L^m f, h) = c^m (f, ([I - k]^{-1} K)^m h),$$

and the functions $\chi_i = ([I - K]^{-1} K)^i h)$ satisfy simultaneous equations (Sect. 5.4):

$$\chi_i = K\chi_i + K\chi_{i-1}, \quad i = 1, ..., m, \chi_0 \equiv h.$$

The corresponding vector algorithm is constructed and investigated as described in Sects. 5.5, 6. The coefficients $a_n^{(i)} = 0$ (see the estimates $Q_n^{(0)}$ of Sect. 5.6) can be calculated in advance using the representation

$$([I - K]^{-1} K)^m = [I - K]^{-m} K^m$$

which holds, since operators K^i and $[I - K]^{-1}$ are commutable. Note that it is convenient to simulate m first collisions without absorption or escape since $a_n^{(i)} = 0$ for $n < i$ [5.7, p.178].

The weight modifications and the correlated sampling in the vector algorithms are realized in exactly the same way as in solving the problems of the transfer theory without fission [5.7]. This algorithm is parallelized automatically. In addition, if the number of the collision type is considered as one of the coordinates of the phase space, then the function $h(x)$ can be concentrated on the fissions, i.e., the quantity $(L^m f, h)$ can be considered to be a functional of the multiplication density for the m-th generation.

5.9 Variance Reduction for the Vector Estimates

In this section we construct vector estimates of a given functional which have minimal variances. The original system of integral equations is written as

$$\varphi_i(x) = \sum_{j=1}^{m} \int_X k_{ij}(x,y)\varphi_i(y)dy + h_i(x), \tag{5.41}$$

or

$$\Phi = K\Phi + H.$$

We assume that all the entries of this system are nonnegative. In this case it is possible to formally construct a Monte Carlo estimate for $\{\varphi_i(x)\}$ with zero variance. To this end, it is sufficient to rewrite (5.41) in the form of a one integral equation (using the corresponding discrete continuous measure), and then to apply the importance sampling principle (Sect. 2.1). The random quantity obtained in such a way is a scalar estimate.

For solving the problem of the variance reduction we will use non-Markov models. Define a chain of random points $x_0 = x, x_1, ..., x_N$ with the transition densities

$$p_{(n-1)n} = p_{(n-1)n}(x_0, ..., x_{n-1}, x),$$

where the quantity $1 - \int p_{(n-1)n}(x_0, ..., x_{n-1}, x)dx$ is the probability that the chain breaks after the transition $x_{n-1} \to x_n$. Next introduce random weights

$$Q_0 = \|\delta_{ij}\|, \quad Q_n = Q_{n-1}\frac{K_{(n-1)n}}{p_{(n-1)n}},$$

where

$$K_{(n-1)n} = K(x_{(n-1)}, x_n) = \|k_{ij}(x_{n-1}, x_n)\|, \quad i,j = 1, ..., m,$$

and the vector random variable

$$\xi_x = \sum_{n-1}^{N} Q_n H_n, \quad H_n = (h_1(x_n), ..., h_m(x_n)).$$

We assume, as in Sect. 5.1, that

$$K \in [L_\infty \to L_\infty], \quad H \in L_\infty, \rho(K_1) < 1,$$

where K_1 is an operator which is obtained from K by replacing $|k_{ij}|$, $i, j = 1, ..., m$.

Then, if $p_{(n-1)n} \neq 0$ when $K_{(n-1)n}\Phi_n \neq 0$, then $\mathbf{M}\xi_x = \Phi(x)$, i.e., ξ_x is an unbounded estimate for the function $\Phi(x)$ [5.8].

Let $\Psi_0 = \Psi(x_0) = \mathbf{M}(\xi_x \xi_x')$. Using iterated averaging, we obtain

$$\Psi_0 = \Psi(x_0) = \mathbf{M}[(H_0 + Q_1\xi_{x_1})(H_0 + Q_1\xi_{x_1})'] = A_0 + \int \frac{K_{01}\Psi_1 K_{01}}{p_{01}} dx_1,$$

where

$$A_0 = H_0\Phi_0' + \Phi_0 H_0' - H_0 H_0', \quad \Psi_1 = \Psi_1(x_0, x_1) = \mathbf{M}(\xi_{x_1}\xi_{x_1}').$$

Note that ξ_{x_1} also depends on x_0, when x_1 is fixed, since the chain is non-Markovian. By induction we get

$$\begin{aligned} \Psi_0 &= A_0 + \int \frac{K_{01}A_1K_{01}}{p_{01}} dx_1 + ... \\ &+ \underbrace{\int ... \int}_{n} \frac{K_{01}...K_{(n-1)n}A_n K'_{(n-1)n}...K'_{01}}{p_{01}...p_{(n-1)n}} dx_1...dx_n \\ &+ \underbrace{\int ... \int}_{n+1} \frac{K_{01}...K_{n(n+1)}\Psi_{n+1}K'_{n(n+1)}...K'_{01}}{p_{01}...p_{n(n+1)}} dx_1...dx_{n+1}, \end{aligned} \tag{5.42}$$

where

$$\begin{aligned} \Psi_{n+1} &= \Psi_{n+1}(x_0, ..., x_{n+1}) = \mathbf{M}(\xi_{x_{n+1}}\xi'_{x_{n+1}}), \\ A_n &= H_n\Phi_n' + \Phi_n H_n' - H_n H_n'. \end{aligned}$$

As in [5.8], we can derive

$$\begin{aligned} \Psi_0 &= A_0 \\ &+ \sum_{n=0}^{\infty} \underbrace{\int ... \int}_{n+1} \frac{K_{01}...K_{n(n+1)}A_{n+1}K'_{n(n+1)}...K'_{01}}{p_{01}...p_{n(n+1)}} dx_1...dx_{n+1}, \end{aligned} \tag{5.43}$$

provided that the last series also converges if we replace k_{ij} with $|k_{ij}|$ and h_i with $|h_i| (i, j = 1, ..., m)$.

Furthermore we consider the problem of optimization of the vector estimate ξ_x which consists of finding a set of transition densities $\{p_{n(n+1)}\}$ such that the variance of the random estimate $l'\xi_x$ is minimal. This means that the quantity

$$l'\Psi_0 l = l'A_0 l + \int \frac{l'K_{01}A_1K_{01}l}{p_{01}}dx_1 + ...$$
$$+ \underbrace{\int ... \int}_{n+1} \frac{l'K_{01}...K_{n(n+1)}\Psi_{n+1}K'_{n(n+1)}...K'_{01}l}{p_{01}...p_{n(n+1)}}dx_1...dx_{n+1}, \tag{5.44}$$

must be minimized. For example, if $l' = (1, 0, ..., 0)$, the problem is to minimize the variance of the first component of ξ_x.

By the Bellmann principle [5.13, 14], the optimal transition density $p_{n(n+1)}$ minimizes the last integral in (5.44), i.e., in the optimal variant we have

$$p_{n(n+1)} = \frac{(l'K_{01}...K_{n(n+1)}\Psi_{n+1}K'_{n(n+1)}...K'_{01}l)^{1/2}}{\int(l'K_{01}...K_{n(n+1)}\Psi_{n+1}K'_{n(n+1)}...K'_{01}l)^{1/2}dx_{n+1}} \ . \tag{5.45}$$

Thus, the following result can be formulated.

Theorem 5.7. For the optimal set of $\{p_{n(n+1)}\}$ the following relations hold:

$$\Psi_n = A_n + \int \frac{K_{n(n+1)}\Psi_{n+1}K'_{n(n+1)}}{p_{n(n+1)}} \ , \tag{5.46}$$

where the functions $p_{n(n+1)}$ are defined by (5.45) at $n = 0, \ldots$. □

The relations (5.45, 46) show that the optimal chain is in general non-Markovian. An exception is the case of one equation $(m+1)$ for which (5.45, 46) takes the form of nonlinear equations

$$\Psi_n = A_n + \left(\int |k_{n(n+1)}|\Psi_{n+1}^{1/2}dx_{n+1}\right)^2 . \tag{5.47}$$

Assume that

$$\text{vrai sup} \int |k(x,y)|dy = q < 1, \quad A_n = A(x_n) \geq 0.$$

Then the operator

$$G(\Psi) = [A + (K_1\Psi^{1/2})^2]^{1/2}$$

is contractive in L_∞; thus the following result is true.

Theorem 5.8. Let $A_n \geq 0, q < 1$. Then there exists a unique solution to (5.47) in L_∞, and

$$\Psi_n(x_0, ..., x_n) = \Psi(x_n), \quad \Psi = A + (K_1\Psi^{1/2})^2,$$

$$p_{n(n+1)} = p(x_n, x_{n+1})$$
$$= k(x_n, x_{n+1})\Psi^{1/2}(x_{n+1}) \left[\int k(x_n, x_{n+1})\Psi^{1/2}(x_{n+1})dx_{n+1}\right]^{-1}.$$

The following theorem is of practical interest. □

Theorem 5.9. Assume that the solutions $\{\Psi_{1n}\}, \{\Psi_{2n}\}$ to (5.45, 46) satisfy the condition

$$\text{vrai sup} \int \frac{\left(\sum_j |k_{ij}(x,y)|\right)^2}{p_{n(n+1)}(x_0, ..., x_{n-1}, x, y)} dy \le q < 1$$

Then $\Psi_{1n}(\cdot) = \Psi_{2n}(\cdot) \quad (n = 0, 1, ...)$.

Proof. We define B_n by (5.44) where $p_{(k+1)k} = p_{1,(k+1)k}, \Psi_{n+1} = \Psi_{2,n+1}$. It is not difficult to show that $B_n \ge l'\Psi_{2,0}l$. The last term for B_n tends to zero by the theorem condition. Therefore, $B_n \to l'\Psi_{1,0}l$ and $l'\Psi_{1,0}l' \ge l'\Psi_{2,0}l$, i.e., $l'\Psi_{1,0}l = l'\Psi_{2,0}l$. Analogously the equality for the functions of type $l'K_{01}...K_{n(n+1)}\Psi_{n+1}k'_{n(n+1)}...K'_{01}l$, which define $\{p_{n(n+1)}\}$, is established; this determines, in view of (5.43), the set $\{\Psi_n\}$. □

Let us now consider the problem of optimization of ξ_x under the condition

$$p_{n(n+1)} = p(x_n, x_{n+1}), \qquad \int p(x,y)dy \equiv q(x).$$

The problem is to determine an optimal stationary Markov chain with the given termination probability. To solve it, we seek a conditional minimum (over $p(\cdot,\cdot)$) of the expression

$$l'\Psi_0 l = l'A_0 l + \sum_{n=0}^{\infty} \underbrace{\int \cdots \int}_{n+1} \frac{1}{p_{01}\cdots p_{n(n+1)}} \times l'K_{01}...K_{n(n+1)}A_{n+1}K'_{n(n+1)}...K'_{01}l \quad dx_1...dx_{n+1}, \tag{5.48}$$

where $p_{n(n+1)} = p(x_n, x_{n+1})(n = 0, 1, ...)$. Let $p^*(x,y)$ be the optimal transition density, and let

$$p_\alpha(x,y) = p^*(x,y) + dr(x,y),$$

where $\int r(x,y)dy \equiv 0$. We replace $p(\cdot,\cdot)$ with $p_\alpha(\cdot,\cdot)$, differentiate (5.48) with respect to α, and invert the order of summation and differentiation. This yields

$$-\sum_{n=0}^{\infty}\sum_{s=0}^{n} \underbrace{\int \cdots \int}_{n+1} \frac{r(x_s, x_{s+1})}{p_\alpha(x_s, x_{s+1})} \times \frac{l'K_{01}...K_{n(n+1)}A_{n+1}K'_{n(n+1)}...K'_{01}l}{p_{\alpha,01}\cdots p_{\alpha,n(n+1)}} dx_1...dx_{n+1}.$$

We also invert the order of summation formally

$$-\sum_{s=0}^{\infty}\underbrace{\int\cdots\int}_{s+1}\frac{r(x_s,x_{s+1})}{p_\alpha(x_s,x_{s+1})}$$
$$\times\frac{l'K_{01}\ldots K_{s(s+1)}\Psi_{s+1}K'_{s(s+1)}\ldots K'_{01}l}{p_{\alpha,01}\cdots p_{\alpha,s(s+1)}}dx_1\ldots dx_{s+1}. \tag{5.49}$$

This expression at $\alpha = 0$ should be equal to zero, because $p^*(\cdot,\cdot)$ is optimal. We now write the corresponding equation

$$\int\int\frac{r(y_1,y_0)}{p^{*2}(y_1,y_0)}dy_0dy_1$$
$$\times\sum_{s=1}^{\infty}\underbrace{\int\cdots\int}_{s-1}\frac{l'K_{(s+1)s}\ldots K_{10}\Psi_0K'_{10}\ldots K'_{(s+1)s}l}{p^*_{(s+1)s}\cdots p^*_{21}}dy_2\ldots dy_s$$
$$+\int\frac{r(y_1,y_0)}{p^{*2}(y_1,y_0)}l'K_{10}\Psi_0K'_{10}l\,dy_0=0,\quad y_{s+1}\equiv x_0\quad(y_1\equiv x_0).$$

The last equation is true for an arbitrary continuous function $r(x,y)\not\equiv 0$ such that $\int r(x,y)dy\equiv 0$. Therefore, for almost all y_1

$$p^*(y_1,y_0)=c(y_1;p^*)$$
$$\times\left[\sum_{s=1}^{\infty}\underbrace{\int\cdots\int}_{s-1}\frac{l'K_{(s+1)s}\ldots K_{10}\Psi_0K'_{10}\ldots K'_{(s+1)s}l}{p^*_{(s+1)s}\cdots p^*_{21}}dy_2\ldots dy_s\right]^{1/2}, \tag{5.50}$$

where $p^*(y_1,y_0)=c(y_1;p^*)(l'K_{01}\Psi_0K'_{10}l)^{1/2},\quad y_1=x_0$, and $c(y_1;p^*)$ is determined from

$$\int p^*(y_1,y_0)dy_0\equiv q(y_1).$$

Let us now give some remarks about the validity of the derivation of (5.49, 50). It is possible to assume that the function $r(x,y)$ is bounded: $|r(x,y)|\leq c(r)$. We also assume that $p^*(x,y)\geq\varepsilon>0$, and

$$\text{vrai sup}\int\frac{\left(\sum_j|k_{ij}(x,y)|\right)^2}{p^*(x,y)}dy=q<1. \tag{5.51}$$

This condition is realistic, since it presents a natural criterion of the finiteness of (5.43) which is formulated in Sect. 5.1.

By this assumption we can define a small value $\alpha_0(r) > 0$ for the arbitrary function $r(x, y)$ such that (5.51) is satisfied if we replace p^* with $p_\alpha = p^* + \alpha r$ where $q = q_1 < 1$ at $|\alpha| < \alpha_0(r)$. Then

$$\left| \int \cdots \int \frac{r(x_s, x_s+1)}{p_\alpha(x_s, x_s+1)} \times \frac{l' K_{01} \ldots K_{n(n+1)} A_{n+1} K'_{n(n+1)} \ldots K'_{01} l}{p_{\alpha,01} \cdots p_{\alpha,n(n+1)}} dx_1 \ldots dx_{n+1} \right| \leq c q_1^{n+1}.$$

Thus the following theorem is justified.

Theorem 5.10. Let $p^*(\cdot,\cdot)$ be an optimal density under the condition

$$\int p(x, y) dy = q(x) \leq 1.$$

Suppose that (5.51) is satisfied and $p^*(x, y) \geq \varepsilon > 0$. Then $p^*(\cdot,\cdot)$ solves (5.50), where Ψ_0 is defined by the expression of type (5.46). □

Notice that when different $q(\cdot)$ are considered, the optimal value is $q(r) \equiv 1$, since from (5.44) it follows that if $p(x, y)$ is replaced with $p(x, y)/q(x)$, the quality $l'\Psi l$ is decreasing. For a positive definite matrix A this follows directly from (5.48).

We now construct an approximation to (5.50) which is more convenient for practical use. Assume that all the terms of the sum in (5.50) are proportional to $(l' K_{10} \Psi_0 K'_{10} l)^{1/2}$; this assumption is true if $m = 1$. As a result, we obtain

$$\begin{aligned} \Psi_0 &= A_0 + \int \frac{K_{01} \Psi_1 K'_{01}}{p_{01}} dx_1, \\ p_{01} &= \frac{(l' K_{01} \Psi_1 K'_{01} l)^{1/2}}{\int (l' K_{01} \Psi_1 K'_{01} l)^{1/2} dx_1} q_0, \end{aligned} \tag{5.52}$$

where $\Psi_i = \Psi(x_i), p_{01} = p(x_0, x_1)$. For $m + 1, \equiv 1$ this yields

$$\Psi = A + (K_1 \Psi^{1/2})^2,$$

which appeared in Theorem 5.8. Thus, (5.52) may be useful if it is necessary to minimize the variance of the most important coordinate of the vector ξ_x. The system (5.52) can solved by the method of successive iterations. The new value of $p(\cdot)$ is calculated on the basis of

$$l' K_{01} \Psi_1 K'_{01} = \mathbf{M}\{(K'_{01} l)' \xi_{x_1}\}.$$

Numerical calculations for a system of four linear algebraic equations written in the form of two integral equations of the second kind

$$\begin{aligned} \varphi_i(j) = {} & k_{i1}(j, 1)\varphi_1(1) + k_{i1}(j, 2)\varphi_1(2) + k_{i2}(j, 1)\varphi_2(1) \\ & + k_{i2}(j, 2)\varphi_2(2) + h_i(j), \quad i, j = 1, 2, \end{aligned}$$

were carried out.

The coefficients of the system were taken as follows (along the row): 0.2, 0.15, 0.1, −0.1, 0.15, −0.1, 0.25, 0.15, 0.15, 0.25, −0.2, 0.1, −0.1, 0.1, 0.1, 0.2 and $h = (3, 1, 2, 1, 1, 1)$. The vector l was taken to be (0,1), i.e., the variances of $\varphi_i(i), i = 1, 2$ were minimized.

A direct resort has given the following optimal value of the transition probabilities: $P^* = (0.76, 0.24, 0.82, 0.18)$ while (5.52) has given $P = (0.78, 0.22, 0.83, 0.17)$. The corresponding values of $\mathbf{M}\xi_x^{(1)2}$ were almost equal. Thus, (5.52) works satisfactorily, although the integral equations are essentially connected. It is interesting to note that one iteration of (5.53) starting from $P^{(0)} = (0.4, 0.6, 0.4, 0.6)$ has given $P^{(1)} = (0.75, 0.25, 0.82, 0.18)$.

5.10 Asymptotic Investigation of a Monte Carlo Method Combined with the Method of Finite Sums

In contrast to Sect. 5.3, we now present the main integral equation in the form

$$\varphi(x) = \int_D k(x', x)\varphi(x')dx' + f(x), \tag{5.53}$$

or

$$\varphi = K\varphi + f.$$

Note that this equation was regarded in the previous sections (and throughout the book) as an adjoint equation in L_1. It is assumed that $\|K\| < 1$.

We define the linear functional for $h \in L_\infty$:

$$I_h = (\varphi, h) = \int_D \varphi(x)h(x)dx.$$

Assume that D is a bounded domain of the l-dimensional Euclidian space. Divide X in m subdomains $D_1, ..., D_m$ such that $D_i \cap D_j = 0$ for $i \neq j$. The Monte Carlo algorithms of this section are based on the representation of (10.1) in the form of the following system of integral equations:

$$\varphi(x) = \sum_{j=1}^{m} \int_{D_j} k(x', x)\varphi(x')dx' + f(x), \quad x \in D_i, \quad i = 1, ..., m.$$

Let $\{X_n\}$ be a chain of random vectors with multi-dimentional coordinates $x_n^{(i)} \in D_i$ defined by the transition density $r_n(X_0, ..., X_{n-1}, X) > 0$, the density of the initial vector X_0 and the probability $p_n(X_0, ..., X_n) < 1$ that the chain terminates after the transition from X_n; N is the random number of the last state. Denote by $r_{ni}(X_0, ..., X_{n-1}, x^{(i)})$ the marginal distribution density of the coordinate $x_n^{(i)}$ of the vector X_n.

By **M** we denote the averaging over the distributions of the chain. As in the case of a system of linear algebraic equations [5.7], it can be shown that

$$\mathbf{M}\sum_{n=0}^{N}\sum_{i=1}^{m} q_n^{(i)}\varphi(x_n^{(i)}) = (f,\varphi) = I_h, \tag{5.54}$$

where

$$q_0^{(i)} = \frac{f(x_0^{(i)})}{r_{0i}(x_0^{(i)})},$$

$$q_0^{(i)} = \sum_{j=1}^{m} q_{n-1}^{(i)} \frac{k(x_{n-1}^{(j)}, x_n^{(i)})}{r_{ni}(X_0,...,X_{n-1},x^{(i)})} \frac{1}{1-p_{n-1}(X_0,...,X_{n-1})}.$$

The Monte Carlo algorithm based on (5.54) can be interpreted as a variant of the method of finite sums where a simple rectangle cubic formula is used, and the system of linear algebraic equations is solved iteratively with a random choice of the cubic nodes $x_n^{(i)}$ in each iteration.

The variance of this estimate probably decreases as m increases, while the computer time required to construct one trajectory is increasing. We consider below the question of optimizing the choice of m.

We suppose that the following conditions are satisfied:

(A) the points $x_n^{(i)} (i = 1,...,m)$ are uniformly distributed in the corresponding domains D_i whose diameters have the order $m^{-1/l}$; $p_n(x) \equiv p_n$;

(B) the functions $f(x)$ and $h(x)$ satisfy the Lipshitz condition with the index α in the domain D, and the function $k(x',x)$ satisfies the same condition in the domain $D \times D$.

We now introduce the following notation: K_{mn} is the integral operator with a piece-wise constant kernel $k_{mn}(x',x) = k(x_{n-1}^{(i)}, x_n^{(j)})$ for $x' \in D_i, x \in D_j$; f_m, h_{mn} are piece-wise constant functions which take in D_i the values $f(x_0^{(i)})$ and $h(x_n^{(i)})$, respectively; $L_{mn} = (K_{mn}...K_{m1}f_m, h_{mn})(n = 1,2,...)$.

Under the above assumption (5.54) can be rewritten as

$$I_h = \mathbf{M}\xi_m,$$

where

$$\xi_m = (f_m, h_{m0}) + \sum_{n=1}^{N}\left(L_{mn}\prod_{k=0}^{n-1}\frac{1}{1-p_k}\right). \tag{5.55}$$

We now estimate the variane $\mathbf{D}\xi_m$. Let us first formulate two statements.

Lemma 5.8. If the conditions (A) and (B) are satisfied, then

$$\|f_m - f\| \le c_0 m^{-\alpha/l},$$

$$\|h_{mn} - h\| \leq c_1 m^{-\alpha/l},$$
$$\|K - K_{mn}\| \leq c_2 m^{-\alpha/l}.$$

The proof is obvious. □

Thus, we will assume that

$$\|K_{mn}\| < q, \quad \|K\| \leq q < 1. \tag{5.56}$$

Lemma 5.9. If the conditions (A), (B) and (5.56) are satisfied, then

$$|L_{mn} - (K^n f, h)| \leq c_3 (n+2) q^{n-1} m^{-\alpha/l}.$$

Proof. The proof follows from simple arguments:

$$L_{mn} - (K^n f, h) = [L_{mn} - (K K_{m(n-1)} \ldots K_{m1} f_m, h_{mn})] +$$
$$+ [(K K_{m(n-1)} \ldots K_{m1} f_m, h_{mn}) - (K^2 K_{m(n-2)} \ldots$$
$$\ldots K_{m1} f_m, h_{mn})] + \ldots + [K^n f, h_{mn}) - (K^n f, h)].$$ □

Now it is not difficult to obtain an upper bound of the variance $\mathbf{D}\xi_m$ provided some additional assumptions about the terminating probability p_n are made.

Theorem 5.11. Let $p_n \equiv 0$ for $n \leq s, 0 < p \leq p_n \leq 1 - q$ for $n > s$. Assume that the conditions (A), (B) are satisfied. Then

$$\mathbf{D}\xi_m \leq c_4 m^{-2\alpha/l} + c_s p^{-2} q^{2s}. \tag{5.57}$$

Proof. Under the conditions of the theorem, the following relation holds:

$$\xi_m = \eta_m + \zeta_m,$$

where

$$\eta_m = \sum_{n=1}^{s} \mathrm{L}_{mn} + (f_m, h_{m0}), \quad \zeta_m = \sum_{n=s+1}^{N} \left(L_{mn} \prod_{k=0}^{n-1} \frac{1}{1-p_k} \right).$$

It follows from Lemma 5.9 that

$$\mathbf{D}\eta_m \leq c_6 \left(\frac{1}{m^{\alpha/l}} \sum_{n=1}^{\infty} (n+2) q^{n-1} + \frac{1}{m^{\alpha/l}} \right)^2 = \frac{c_4}{2m^{2\alpha/l}}.$$

From (5.56) we obtain

$$|\zeta_m| \leq c_7 q^s (N - s), \quad N \geq s+1 \quad ,$$

since $0 < p < p_n \leq 1 - q$.

The random variable $N - s$ has a geometric distribution. Therefore,

$$\mathbf{D}\xi_m \le \mathbf{M}\zeta_m^2 \le q^{2s}\frac{c_5}{2p^2}.$$

This completes the proof, since

$$\mathbf{D}\xi_m \le 2(\mathbf{D}\eta_m + \mathbf{D}\zeta_m)$$ □

Simple arguments show that the order of the estimatation of $\mathbf{D}\xi_m$ cannot be improved. Thus, we will assume in what follows that $\mathbf{D}\xi_m$ has an order equal to that of the quantity in the right-hand side of (5.57). The effectiveness of an algorithm is defined as the computer time required to achieve a given accuracy. In our case the effectiveness is proportional to $\Delta_m = t_m \mathbf{D}\xi_m$, where t_m is the time of simulation of the trajectory $\{X_n\}$. It is clear that $t_m = O(m^2)$.

Let $p_n \equiv p \le 1 - q$ for $n \ge s$. Then

$$\Delta_m \approx c_8 \left(\frac{1}{p} + s\right)\left(\frac{c_4}{2m^{2\alpha/l}} + c_s \frac{q^{2s}}{p^2}\right) m^2. \tag{5.58}$$

Note that this expression is of minimal order if $q^{2s} \approx m^{-2\alpha/l}$, i.e., when $s \approx \alpha \ln m / |l \ln q|$. Substituting this into (5.58) yields

$$\Delta_m \approx c_9 m^{2(1-\alpha/l)} \ln m \to \infty, m \to \infty,$$

since the value $\alpha > 1$ may appear only for a function which is equivalent to a constant. Therefore, it is useful to combine the Monte Carlo method with the method of finite sums for calculating one functional for an arbitrary value $\alpha \le 1$.

We describe briefly the question on how to evaluate the solution in many points (i.e., to calculate the whole solution field). Assume that the solution is calculated by the histogram, as is usual for calculating a probability density [5.15]. In [5.15] it is shown that the mean squared probabilistic error of the optimal histogram approximation under some additional assumptions is equal to $\sigma^{2+l}t$, where t is the computer time for one trajectory.

It can be shown analogously that the error of the piece-wise constant approximation of the solution to (5.53) has the order $m^{-\alpha/l}$, when a single trajectory is used, and if

$$s = \alpha \ln m / |l \ln q|.$$

Therefore, for this choice of s we get

$$\sigma = \sigma_m \lesssim c_{10} m^{-\alpha/l},$$

$$\Delta_m^{(1)} = \sigma_m^{2+l} t_m \lesssim c_{11} m^{2-\alpha(2+l)/l} \ln m.$$

Thus if $\alpha > 2l/(2+l)$, then $\Delta_m^{(1)} \to 0$ as $m \to \infty$. This implies that it is not useful to apply the Monte Carlo method in this case. So the strategy when the number of nodes in the method of finite sums is increased is preferable compared to the method which uses a set of trajectories $\{X_n\}$. Note, however, that the inequality $\alpha > 2l/(2+l)$ is possible only if $l \equiv 1$.

6. Randomization of Weighted Algorithms

Various examples introducing additional randomness for constructing effective simulation algorithms can be found in the literature devoted to the Monte Carlo methods. This chapter is concerned with the optimization of randomized algorithms for estimating probabilistic characteristics of equations with random parameters. In this connection, randomized models for random fields are suggested. A technique is presented for approximate averaging of the exponential factors of the kernels of the transfer integral equations.

6.1 Randomized Estimation for Statistical Moments of the Solution

Assume a linear functional equation $L\varphi = f$ is to be solved by the Monte Carlo method on the basis of simulation of a stochastic process. (Denote the trajectories of this process by ω). This means that random variables $\xi_k(\omega)$ are constructed so that

$$\mathbf{M}\xi_k(\omega) = I_k, \quad k = 1, \dots, m$$

where I_k are the functionals of φ to be evaluated ($\mathbf{M}$ denotes the mathematical expectation).

Let the operator L and the function f depend on a random field σ (e.g., a random medium in transfer theory, random force in elasticity theory, etc.). Also, $\xi_k = \xi_k(\omega, \sigma), I_k = I_k(\sigma)$, and $\mathbf{M}[\xi_k(\omega, \sigma|\sigma] = I_k(\sigma)$, where the variables ω and σ are generally not dependent.

Let us consider the problem of evaluating the quantities

$$I_k = \mathbf{E}I_k(\sigma), \quad R_{kj} = \mathbf{E}\{I_k(\sigma)I_j(\sigma)\}, \quad k, j = 1, \dots, m,$$

where $\mathbf{E}$ denotes the mathematical expectation with respect to the distribution of σ. The following obvious method is known for evaluating these mathematical expectations. First, realizations of σ are constructed; then the equation $L_\sigma \varphi_\sigma = f_\sigma$ is solved precisely enough for each realization by a numerical or an analytical technique. Finally, statistical estimates of the desired quantities are calculated.

However, this approach fails for complicated multi-dimensional problems because the computational cost of an explicit solution of the equation considered is too high. Therefore, it is useful to sometimes apply a method of double randomization. In our case, this technique follows from

$$\mathbf{E}I_k(\sigma) = \mathbf{E}\mathbf{M}\xi_k(\omega,\sigma) = \mathbf{M}_{(\omega,\sigma)}\xi_k(\omega,\sigma),$$

$$\mathbf{E}\{I_k(\sigma)I_j(\sigma)\} = \mathbf{M}_{(\omega_1,\omega_2,\sigma)}\{\xi_k(\omega_1,\sigma)\xi_j(\omega_2,\sigma)\}, \tag{6.1}$$

where ω_1 and ω_2 are conditionally independent trajectories, constructed for one fixed realization of σ, and the subscript of the expectation symbol indicates the distribution to which it corresponds. Relations (6.1) show that in order to estimate the quantities I_k, it is sufficient to construct only one trajectory for a fixed σ, while the estimation of the quantities R_{kj} requires two conditionally independent trajectories. To optimize the randomization technique, it is natural to use the splitting method (Sect. 1.5), where, the quantities I_k are estimated as follows. First, one constructs n conditionally independent trajectories (i.e., a vector $\omega = (\omega_1, \dots, \omega_n)$, with σ fixed), and then a random variable

$$\zeta_k(\omega,\sigma) = \frac{1}{n}\sum_{i=1}^{n} \xi_k(\omega_i,\sigma)$$

is used rather than $\xi_k^{(n)}(\omega,\sigma)$. The optimal value of n is calculated by (Sect. 1.5)

$$n = \sqrt{a_2\, t_1 / \, a_1\, t_2}, \tag{6.2}$$

where $a_1 = \mathbf{E}\mathbf{M}_\omega^2\, \xi_k - I_k^2$; $a_2 = \mathbf{E}D_\omega\xi_k$; t_1 is the average computing time for a fixed realization of σ, and t_2 is the average computing time for a fixed realization of ω.

It is difficult to evaluate the quantities a_1, a_2, t_1, t_2 directly. However, one can obtain statistical estimates of the variances and the computing times for two values of the splitting parameter , n_1, n_2, and then solve the corresponding system of linear equations, i.e., make use of the following equalities:

$$\begin{aligned} a_1 &= \frac{1}{n_2 - n_1}(n_2 D\zeta_k^{(n_2)} - n_1 D\zeta_k^{(n_1)}),\\ a_2 &= \frac{n_1 n_2}{n_2 - n_1}(D\zeta_k^{(n_1)} - D\zeta_k^{(n_2)}),\\ t_1 &= \frac{n_2 t^{(n_1)} - n_1 t^{(n_2)}}{n_2 - n_1}, \quad t_2 = \frac{t^{(n_1)} - t^{(n_2)}}{n_2 - n_1}. \end{aligned} \tag{6.3}$$

Here, it is useful to correlate the samples of $\zeta_k^{(n_1)}$ and $\zeta_k^{(n_2)}$.

Let us now consider the optimization of the randomized estimate of the quantity R_{kj} using the splitting technique. In this case, it is natural to use the random variable

$$\rho_{kj}^{(n)} = \frac{1}{n(n-1)/2}\sum_{i=1}^{n}\sum_{t=i+1}^{n} \xi_k(\omega_i,\sigma)\xi_j(\omega_t,\sigma), \quad \mathbf{M}\rho_{kj}^{(n)} = R_{kj}.$$

Note that rather than $\rho_{kj}^{(n)}$ it is possible to use another quantity which is numerically equivalent to $\rho_{kj}^{(n)}$, but more convenient for calculations. Indeed, the following easily verified relation for the covariances holds:

$$K[\zeta_k^{(n)}(\omega,\sigma),\zeta_j^{(n)}(\omega,\sigma)] = K[I_k(\sigma),I_j(\sigma)] + \frac{1}{n}\mathbf{E}K[\xi_k,\xi_j].$$

Taking into account the analogous relation for $n = 1$, it follows from this system of equations that

$$K_{jk} = K[I_k(\sigma),I_j(\sigma)] = \frac{nK[\zeta_k^{(n)},\zeta_j^{(n)}] - K[\zeta_k^{(1)},\zeta_j^{(1)}]}{n-1}. \tag{6.4}$$

It is not difficult to verify that substituting the statistical estimates for the appropriate covariances into the right-hand side of (6.4) yields an estimate $\tilde{K}_{jk}^{(n)}$ of the quantity K_{jk} which coincides numerically with the estimate obtained on the basis of $\rho_{kj}^{(n)}$.

When (6.4) is used, the average computing time is given, as in the standard splitting technique, by the formula

$$T^{(n)} = T_1 + nT_2.$$

The explicit formula for the variance $D\rho_{kj}^{(n)}$ is very cumbersome. However, it is clear that for sufficiently large n the following approximate relation holds:

$$D\tilde{K}_{kj}^{(n)} \approx A_1 + \frac{A_2}{n}.$$

Consequently, the splitting technique, when used for estimating K_{jk} , can also be optimized by a formula of type (6.2, 3).

It should be noted that we do not consider the optimization of the choice (in some sense) of n for estimating all the quantities K_{jk}, $j,k = 1,\dots,m$, simultaneously.

We now assume that the random field σ is entirely defined by the finite number of the random parameters $\sigma_1,\dots,\sigma_r$. In addition, we assume the dependence of the functionals $I_k(\sigma)$ on σ_i, $i = 1,\dots,r$ to be sufficiently linear (this assumption can be made in the neighbourhood of the point $\sigma^{(0)} = \mathbf{M}\sigma$; the probability to hit this neighbourhood is close to 1).

Then the correlations of the vector $I = \{I_k(\sigma)\}$ and the autocorrelations of the vectors σ and I can be evaluated by linearizing $I(\sigma)$, using Monte Carlo calculations of the estimates of the derivatives $\partial I_k/\partial\sigma_1$ (Sect.1.8). The corresponding numerical technique is given in the transfer theory for piece-wise constant random cross section σ. Consider optimization of the double randomization method for calculating functionals of the form

$$I = \mathbf{E}I(\sigma) = \mathbf{EM}\xi(\omega,\sigma)$$

considering the cost of the approximate simulation of the field σ.

Let σ_m be an approximate model for the field σ (Sect. 6.3, 4) and assume that

$$(I_m - I)^2 \lesssim m^{-\alpha}, \quad I_m = \mathbf{E}I(\sigma_m), \quad \alpha > 0.$$

Here, the symbol $\lesssim$ means that the estimation is sufficiently exact. Inequalities of this type can be obtained using *a priori* information of the solution on the basis of estimates of the corresponding norms of the random function $\sigma - \sigma_m$.

Assume also the variance

$$d = D\xi(\omega, \sigma_m) = \mathbf{EM}[\xi(\omega, \sigma_m) - I_m]^2$$

to be independent of m. The probabilistic error of the estimate for $I_m(\omega, \sigma_m)$ is defined by the quantity

$$\mathbf{EM}[\xi(\omega, \sigma_m) - I]^2 = d + (I_m - I)^2.$$

For $\xi_N(\omega, \sigma_m)$, i.e., for the average value of N independent realization of $\xi(\omega, \sigma_m)$, we have

$$\mathbf{EM}[\xi_N(\omega, \sigma_m) - I]^2 \lesssim \frac{d}{N} + cm^{-\alpha}.$$

Let us measure the cost of an algorithm by the average number of operations required to obtain the result to within probabilistic error ε. Assume the average cost of simulation of $\xi(\omega, \sigma)$ for $\sigma = \sigma_m$ to be given by $c_0 m^\beta$, $\beta > 0$. Note that in the algorithms of Sects. 6.3, 4, it is apparently equal to 1. To optimize the randomized algorithm of estimation of the quantity I, it is necessary to solve the following conditional minimization problem:

$$\min_{N,m}(Nm^\beta), \quad d/N + cm^{-\alpha} = \varepsilon^2. \tag{6.5}$$

From (6.5), we have

$$m = [c^{-1}(\varepsilon^2 - d/N)]^{-1/\alpha},$$

i.e., it is necessary to minimize the quantity

$$N(\varepsilon^2 - d/N)^{-\beta/\alpha} = f(N).$$

The solution of the equation $f'(N) = 0$ has the form

$$N = (d + \beta/\alpha)\varepsilon^{-2},$$

or

$$\varepsilon^2 = \frac{d + \beta/\alpha}{N}.$$

From (6.5), we now have

$$\frac{d}{N} + cm^{-\alpha} = \frac{d + \beta/\alpha}{N},$$

hence,

$$N = \frac{\beta}{\alpha c} m^\alpha. \tag{6.6}$$

The main difficulty here when solving a concrete problem is obtaining sufficiently exact values for c and α. It is interesting to note that (6.6) holds also if the splitting technique is used, i.e., when $\xi(\omega, \sigma_m)$ is replaced by $\zeta^{(n)}(\omega, \sigma)$, provided that the average cost of one sampling of $\zeta^{(n)}(\omega, \sigma)$ for $\sigma = \sigma_m$ is also proportional to m^β.

6.2 Lower Bound of the Variance. Averaging Exponential Kernels

We consider an integral equation

$$\varphi(x;\sigma) = \int\limits_X k(x,x';\sigma)\varphi(x';\sigma)\,dx' + h(x;\sigma) \tag{6.7}$$

or

$$\varphi_\sigma = K^{(\sigma)}\varphi_\sigma + h_\sigma,$$

where σ is a random field. It is assumed that

$$\rho(K^{(\sigma)}) \le \rho_0 < 1 \quad \forall\sigma.$$

We denote the corresponding collision estimate by $\xi_x^{(\sigma)}$. Assume that it is necessary to calculate the quantity

$$\mathbf{E}\varphi(x;\sigma) = \mathbf{EM}\xi_x^{(\sigma)}.$$

To solve this problem we consider the randomized (Sect. 6.1) estimate $\xi_x^{(\sigma)}$ which is constructed for different values of σ on one and the same Markov chain with the transition density $p(x,x')$. The variance of such an estimate has the form

$$\begin{aligned}\Psi(x) &= \mathbf{EM}\xi_x^{(\sigma)2} \\ &= \mathbf{E}\chi(x;\sigma) + \int\limits_X \frac{\mathbf{E}\{k^2(x,x';\sigma)\mathbf{M}\xi_{x'}^{(\sigma)2}\}}{p(x,x')}\,dx'\ .\end{aligned}$$

Using the Bellmann principle, it is possible, as in Sect. 5.9, to construct the optimal non-Markovian transition density. In addition, it is possible to construct formal relations which define the optimal Markov chain. However, it is difficult to use this technique in practice. Therefore, we derive only an equation which defines the lower bound for $\Psi(x)$. Heuristic arguments show that the lower bound can give a good approximation of $\Psi(x)$ in some stochastic problems of transfer theory.

Henceforth, we use the notation $\mathbf{E}\chi(x,\sigma) = \chi_0(x)$. Obviously,

$$\Psi(x) \geq \chi_0(x) + \int_X \frac{\mathbf{E}k^2(x,x';\sigma)\Psi(x')}{p(x,x')}\,dx'.$$

Denote by $K_p^{(0)}$ the integral operator with the kernel

$$\frac{k_0^2(x,x')}{p(x,x')} = \frac{\mathbf{E}k^2(x,x';\sigma)}{p(x,x')}.$$

Suppose that $\rho(K_p^{(0)}) < 1$. Then by Lemma 1.1 we have $\Psi(x) \geq \Psi_0(x)$, where Ψ_0 is the solution to the equation

$$\Psi_0(x) = \chi_0(x) + \int_X \frac{k_0^2(x,x')}{p(x,x')}\,\Psi_0(x')\,dx'.$$

The quantity $\Psi_0(x)$ is finite if $p(x,x') \equiv k_0(x,x')$ and if

$$\int_X k_0(x,x')\,dx' \leq q_0 < 1.$$

The estimate constructed can be minimized using the Theorem 2.3 taking $\rho_1^2 = \chi_0$, $k(x,x') = k_0(x,x')$. This leads to an effective Monte Carlo algorithm if $\Psi \approx \Psi_0$. Note that replacing $\mathbf{E}k^2$ by the product of mean values of the co-factors of the kernel k^2 results in strengthening the estimate for Ψ mentioned above. From this, it is not difficult to obtain the following inequality for the integral equation of radiative transfer in a stochastic medium:

$$\mathbf{EM}\xi_x^{(\sigma)2} \geq \mathbf{M}\xi_x^2,$$

where ξ_x is a random estimate for the solution of the integral equation of transfer in a "mean medium", i.e., in a medium with a mean scattering cross section, a mean attenuation and a mean scattering indicatrix. To obtain this inequality, the following property was used:

$$\mathbf{E}\exp\{-2\tau\} \geq \exp\{-2\mathbf{E}\tau\}.$$

A more accurate estimate can be derived using the exact value of $\mathbf{E}\exp\{-2\tau\}$. We now assume the medium to be stochastically homogeneous. The form of the kernel of the transfer integral equation shows that the quantity $\mathbf{M}\xi_x^{(\sigma 2}$ and the co-factors of the kernel

$$\exp\{-2\tau\}, \quad G(\mu,r') = \sigma_s(r')w(\mu_s,r')$$

are weakly dependent if the correlation scale of the medium is far less than the mean free path length of the particle.

Therefore, Ψ_0 is a good approximation of Ψ if the kernel in (6.7) is constructed as the product of $\mathbf{E}\exp\{-2\tau\}$ and $\mathbf{E}G^2(\mu_s,r')$. In addition, this kernel is close to the kernel of the operator K_p for a deterministic homogeneous medium with the characteristics

$$\sigma_s = \int [\mathbf{E}G^2(\mu, r')]^{1/2}\, d\mu, \quad w_s(\mu_s) = [\mathbf{E}G^2(\mu_s, r')]^{1/2}\sigma_s^{-1} \tag{6.8}$$

provided that the following relation holds:

$$\mathbf{E}\exp\{-2\tau(l)\} \approx \exp\{-2\sigma_0 l\} \quad ,$$

where the total extinction coefficient is σ_0. If $\sigma_s/\sigma_0 < 1$, then the kernel $k_1(x, x')$ of the corresponding transfer equation is substochastic (Sect. 1.6) and the quantity Ψ is likely to be finite if $p(x, x') = k_1(x, x')$.

The arguments considered show that when only one Markov chain is used for estimating $\xi_x^{(\sigma)}$, the appropriate choice of $\{x_n\}$ coincides with the physical process of radiative transfer in the deterministic medium with the mean squared characteristics rather than the mean characteristics.

The practical criterion of variance finiteness is the inequality $\sigma_s/\sigma_0 < 1$, where σ_s is defined from (6.8).

Let us consider a method for calculating the quantity $\mathbf{E}\exp(-\tau(z))$ for a specific Poisson model of the stochastic medium:

$$\tau(z) = \int_0^z \sigma(z')\, dz'.$$

It is assumed that the realizations of the random function $\sigma(z)$ are constructed as follows. A Poisson sequence of points $z_0 = 0$, z_1, $z_2, \ldots$ is constucted such that the random variables

$$\eta_i = z_i - z_{i-1}, \quad i = 1, 2, \ldots,$$

are independently distributed with the density $\lambda\exp(-\lambda t)$, $t > 0$, and an independent value of σ is sampled in each interval (z_{i-1}, z_i) according to a given one-dimensional distribution of this quantity. The constructed random function $\sigma(z)$ is homogeneous with the given one-dimensional distribution and the standardized correlation function is equal to $\exp(-\lambda z)$ (Sect. 6.3). This model is Markovian since the jump over the given level z for the exponential distribution is also exponential. It is suitable for media which do not contain too compactly packed random heterogeneities.

In Sect. 6.3, we show that the correlation function is given by

$$K(z) = \int_0^\infty \exp(-\lambda z)\, \mu(d\lambda) \tag{6.9}$$

if the quantity λ in the above model is sampled from the distribution $\mu(d\lambda)$. Note that (6.9) defines an absolutely monotonic function but the Markov property is then broken. This model of $\sigma(z)$ can be called the generalized Poisson model. We use the additional notation $I(z, \sigma) = \exp\{-\tau(z)\}$.

We first note that in the Poisson model of $\sigma(z)$, the conditional mathematical expectation of the random variable $I(z, \sigma)$, with z_1 fixed, is given by

$$\mathbf{E}I(z,\sigma|z_1) = \begin{cases} \mathbf{E}\exp(-\sigma_1 z_1)\mathbf{E}I(z-z_1,\sigma), & z_1 \le z, \\ \mathbf{E}\exp(-\sigma_1 z), & z_1 > z, \end{cases}$$

where σ_1 is a random value of σ in the interval (z_0, z_1). Averaging this expression over z_1 yields

$$\begin{aligned} \mathbf{E}I(z,\sigma) = \lambda \int_0^z \mathbf{E}\exp(-\sigma_1 z_1)\exp(-\lambda z_1)\mathbf{E}I(z-z_1,\sigma)dz_1 \\ + \exp(-\lambda z)\mathbf{E}\exp(-\sigma_1 z). \end{aligned} \tag{6.10}$$

This is a special case of the so-called *improper* renewal equation. The asymptotic solution to this equation as $z \to \infty$ is given by [6.1]:

$$\begin{aligned} \mathbf{E}I(z,\sigma) &\approx \frac{1}{\mu}\exp(-\alpha z)\int_0^\infty \exp(\alpha t)\exp(-\lambda t)\mathbf{E}\exp(-\sigma_1 t)\,dt \\ &= \frac{1}{\mu}\mathbf{E}(\lambda-\alpha+\sigma_1)^{-1}\exp(-\alpha z), \end{aligned} \tag{6.11}$$

where α is the unique solution of the equation

$$\Phi(\alpha) = \lambda \int_0^\infty \mathbf{E}\exp(-\sigma_1 t)\exp(\alpha t)\exp(-\lambda t)\,dt = 1, \tag{6.12}$$

and the quantity μ is given by

$$\mu = \lambda \int_0^\infty t\mathbf{E}\exp(-\sigma_1 t)\exp(\alpha t)\exp(-\lambda t)\,dt = \lambda\mathbf{E}(\lambda-\alpha+\sigma_1)^{-2}.$$

Equation (6.12) can be rewritten as follows:

$$\lambda\mathbf{E}(\lambda-\alpha+\sigma_1)^{-1} = 1. \tag{6.13}$$

The asymptotics of the mean intensity for the Poisson model of $\sigma(z)$ can thus be found from

$$\mathbf{E}I(z,\sigma) \approx \lambda^{-1}[\mathbf{E}(\lambda-\alpha+\sigma_1)^{-1}\mathbf{E}(\lambda-\alpha+\sigma_1)^{-2}]\exp(-\alpha z) = I_0(z) \quad , \tag{6.14}$$

where α is determined from (6.13). Obviously, $\Phi(\alpha)$ increases monotonically and $\Phi(0) < 1$. Therefore, (6.13) has a unique solution in $(0,\infty)$. To estimate the quantity $\mathbf{E}\exp(-\sigma_1 t)$ in (6.12), we use the Jensen inequality. By averaging (6.14) over λ with respect to $\mu(d\lambda)$, the asymptotics of the function $\mathbf{E}I(z;\sigma)$ can also be found from the obtained results for the generalized Poisson model of $\sigma(z)$ with the correlation function (6.9).

Calculations carried out in [6.2] show that , as a rule, asymptotics (6.14) gives a good approximation to $\mathbf{E}\exp(-\tau(z))$. Note also that this is a lower approximation, since $\mathbf{E}\exp(-\tau(0)) = 1$ while

$$I_0(0) = \frac{\mathbf{E}(\lambda - \alpha + \sigma_1)^{-1}}{\lambda \mathbf{E}(\lambda - \alpha + \sigma_1)^{-2}} = \frac{[\mathbf{E}(\lambda - \alpha + \sigma_1)^{-1}]^2}{\mathbf{E}(\lambda - \alpha + \sigma_1)^{-2}} \leq 1 \quad ,$$

because of (6.13).

6.3 Special Models of Non-Gaussian Random Fields Related to Stationary Point Fluxes

In order to solve stochastic problems by the method of statistical modeling it is necessary to numerically construct the realizations of the random processes and fields. It is then often important to preserve the correlation function and the one-dimensional distributions, even more so because, in the case of non–Gaussian fields, there is no satisfactory information about multi-dimensional distributions. However, it is often desirable to have a set of models of random fields; on the basis of this set, it is possible to study the sensitivity of functionals to the variations of multi-dimensional distributions when they converge to distributions of a random field with appropriate (e.g., continuous) realizations. Models of this kind which are sufficiently close (in the sense of weak convergence) to the limit fields and realizable on a computer are naturally used as approximate models. The more accurate these approximations, the larger is the computational cost of realization.

In this section, we consider special random processes which are connected with stationary fluxes of points

$$\tau_k = \sum_{i=1}^{k} \eta_i, \qquad \tau_0 = 0,$$

where $\{\eta_i\}$ are independent nonnegative random variables with distribution densities $f_i(x)$, $(i = 1, \ldots)$. These processes are called Palm fluxes and the following relations hold:

$$f_k(x) = f(x) = F'(x), \quad k = 2, 3, \ldots, f_1(x) = \mu^{-1}[1 - F(x)]. \tag{6.15}$$

It is assumed that

$$\mu = \int x f(x)\, dx < \infty.$$

On the other hand, it follows from (6.15) that the flux $\{\tau_k\}$ is stationary [6.1]. We use Palm fluxes with a given probability

$$p_0(t) = \mathbf{P}(k = 0; t)$$

on the interval $[0, t]$. The Palm flux is known to be fully defined by $p_0(t)$ [6.3] such that

$$f(x) = -p_0''(x)/p_0'(0), \quad f_1(x) = -p_0'(x), \ x \geq 0. \tag{6.16}$$

It follows directly from (6.15) that

$$p_0(t) = \frac{1}{\mu}\int_t^\infty [1 - F(t')]\,dt'.$$

Therefore,

$$p''(t) \geq 0,\ |p_0'(0) < \infty,\ p_0(0) = 1,\ p_0(\infty) = 0. \tag{6.17}$$

It is not difficult to show that (6.17) implies inequality $p_0' \leq 0$.

Suppose that a given nonnegative function $p_0(t), t \geq 0$ satisfies the condition (6.17). It is then not difficult to verify that (6.16) are the distribution densities for which the second condition in (3.15) holds.

Lemma 6.1. Let $p_0(t)$ be a function such that

$$p_0(0) = 1,\ p_0(\infty) = 0,\ |p_0'(0)| < \infty$$

and $p''(t) \geq 0$ for $t \geq 0$. If (6.15, 16) hold for the Palm flux, then

$$P(k = 0; t) = p_0(t).$$

Using the Palm flux $\{\tau_k\}$ with $P(k = 0; t) = p_0(t)$, we now construct a stationary random process $\xi(t)$, $0 < t < T$ with a given one-dimensional distribution $F_\xi(x)$ as follows:

(1) A sequence $\{\tau_k\}$ is simulated until the first exit outside of $(0, T)$.

(2) One takes $\xi(t) \equiv \xi_j$ in each interval (τ_{i-1}, τ_i), $i = 1, \ldots,$ where $\{\xi_i\}$ are independent random variables with the distribution function $F_\xi(x)$. We denote this process by $\xi(t; p_0)$. The standardized correlation function of the process $\xi(t; p_0)$ is equal to $p_0(t)$, since

$$\mathbf{M}\{\xi(t'; p_0)\xi(t' + t; p_0)\} = p_0(t)\mathbf{M}\xi^2 + [1 - p_0(t)](\mathbf{M}\xi)^2.$$

It is assumed that $\mathbf{M}\xi^2 < \infty$. Thus, we have constructed a stationary process with a given one-dimensional distribution function $F_\xi(x)$ and a convex standardized correlation function $k(t) = p_0(t)$. Note that in accordance with the standard method of the inverse distribution function [6.4], the realization of $\{\tau_k\}$ can be constructed by solving the equations

$$k(\eta_1) = \alpha_1, \quad k'(\eta_i) = \alpha_i k'(0), \quad i = 2, 3, \ldots,$$

where $\{\alpha_i\}$ are independent random variables uniformly distributed on $(0, 1)$.

Consider the case of a monotonic function $k(t)$ which can be represented by

$$k(t) = \int_0^\infty \exp(-\lambda t)\mu(d\lambda) \quad ,$$

where μ is a probabilistic measure on $(0, \infty)$ such that $P(\lambda = 0) = 0$ and

$$\int \lambda\, \mu(d\lambda) < \infty.$$

In this case, (6.17) holds for $p_0(t) = k(t)$ and

$$f(x) = \int_0^\infty \lambda^2 \exp(-\lambda x)\mu(d\lambda) \left[\int_0^\infty \lambda\mu(d\lambda)\right]^{-1}, \ f_1(x) = \int_0^\infty \lambda \exp(-\lambda x)\mu(d\lambda).$$

If convenient methods of sampling the measures μ and $\lambda\mu$ exist, then it is useful to apply the superposition method to sample $\{\eta_i\}$ [6.4] .

Note that there exists a randomized method for constructing a process with an absolutely monotonic correlation function $k(t)$. Indeed, one samples λ with respect to μ and then evaluates $\xi(t; e^{-\lambda t})$.

The process constructed in this manner is obviously not ergodic while the process $\xi(t; k)$ considered above is ergodic and therefore more natural and important for applications.

Let D be a domain of $n-$dimensional Euclidian space lying in a cube $[0, T]^n$. Construct a random field $\xi(r)$, $r = (x_1, \ldots, x_n)$ in D as follows:

(1) Sample a Palm flux $\{\tau_k^{(i)}\}$ with $p_0^{(i)}(t) = k_i(t)$ on the i-th axis.

(2) Sample an independent value of $\xi(r)$ according to the distribution function $F_\xi(x)$ for each parallelepiped of the form

$$[\tau_{k_1-1}^{(1)}, \tau_{k_1}^{(1)}] \times \cdots \times [\tau_{k_n-1}^{(n)}, \tau_{k_n}^{(n)}].$$

Denote this field by $\xi(r; k)$, where $k = (k_1, \ldots, k_n)$. The field $\xi(r; k)$ is homogeneous and has a standardized function

$$K(r; k) = \prod_{i=1}^{n} k_i(|x_i|), \quad r = (x_1, \ldots, x_n). \tag{6.18}$$

Note that the set of simulated correlation functions can be extended, replacing x_i in (6.18) by $\lambda_i x_i$, where $\lambda = (\lambda_1, \ldots, \lambda_n)$ is chosen at random according to the probabilistic measure $\mu(d\lambda)$ defined on $[0, \infty]^n$ for every realization of the field. In this manner it is possible to simulate fields with the correlation function

$$K(r) = \int k_1(|\omega_1|t) \ldots k_n(|\omega_n|t)\, d\omega.$$

The simplest model of this kind is constructed as follows:

(1) A Palm flux $\{\tau_k\}$ with $p_0(t) = k(t)$ is sampled along an isotropic direction ω inside the domain D.

(2) An independent value of ξ is sampled according to $F_\xi(x)$ for each layer between parallel planes, perpedicular to the direction ω, constructed at points $\{\tau_k\}$.

We call such a field model an isotropic-layer model and have

$$K(r) = c_n \int_0^{\pi/2} k(r \cos v) \sin^{n-2} v \, dv.$$

For example,

$$K(r) = \int_0^1 \int_\Lambda e^{-\lambda r \nu} \mu(d\lambda) \, d\nu = \int_\Lambda \frac{1 - e^{-\lambda r}}{\lambda r} \mu(d\lambda) \tag{6.19}$$

if k is absolutely monotonic and $n = 3$. Distribution of the isotropic-layer field is not natural in some applications; however, it can sometimes be improved by the simple modification considered below.

Assume the distribution of ξ to be infinitely divisible, i.e., the representation

$$\xi = \xi_1^{(m)} + \ldots + \xi_m^{(m)}$$

holds for every positive integer m, where $\xi_i^{(m)}$ are independently and identically distributed according to the distribution function $F_m(x)$.

Consider the process

$$\zeta_m(t) = \sum_{i=1}^{m} \xi_i^{(m)}(t; k) \quad , \tag{6.20}$$

where $\xi_i^{(m)}(t; k)$ are independent realizations of the process $\xi^{(m)}(t; k)$ distributed according to the one-dimensional distribution $F_m(x)$. Obviously, the one-dimensional distribution function of $\zeta_m(t)$ is equal to $F_\xi(x)$ and the correlation function is equal to $k(t)$.

The realizations of the process $\zeta_m(t)$ are improved [as compared to $\xi(t; k)$] in the sense that they are closer to continuous functions, because $\zeta_m(t)$ takes constant values in smaller domains and these values are dependent.

Let us now investigate the weak convergence of the processes $\zeta_m(t)$. First it is necessary to derive the convergence of the corresponding marginal distributions. For Gaussian distribution function $F_\xi(x)$ this convergence follows from the Central Limit Theorem.

To derive the limit distributions in the general case, we note that each realization of the flux of points divides the fixed sequence of points $T_k = (t_1, \ldots, t_k)$ into sub-sequences of length $k_1^{(s)}, \ldots, k_{n_s}^{(s)}$ where s is the index of the subdivision. The total number of all subdivisions is equal to 2^{k-1}. We denote the probability of the appearance of the s-th subdivision by p_s.

Lemma 6.2. Let $\varphi_{km}(T_k; u_1, \ldots, u_k)$ be the characteristic function of k-dimensional marginal distribution of the process $\xi^{(m)}(t; k)$. Then

$$(1) \quad \varphi_{km}(T_k; u_1, \ldots, u_k) \to \prod_{s=1}^{2^{k-1}} \left[\prod_{j=1}^{n_s} \varphi(u_{q(s,j)} + \ldots + u_{q(s,j+1)-1}) \right]^{p_s} ,$$
$$= \psi_k(T_k; u_1, \ldots, u_k)$$

as $m \to \infty$, and

$$q(s,j) = \sum_{i=1}^{j-1} k_i^{(s)} + 1, \quad q(s,1) = 1;$$

(2) ψ_k are the characteristic functions of distributions satisfying the consistency conditions.

Proof. We give the detailed proof for the case $k = 2$. Let $\xi_i = \xi^{(m)}(t_i; k)$ and $\Delta t = |t_2 - t_1|$. Then

$$\begin{aligned}\varphi_{2m}(T_2; u_1, u_2) &= \mathbf{M} \exp\{i(\xi_1 u_1 + \xi_2 u_2)\} \\ &= p_0(\Delta t)\varphi^{1/m}(u_1 + u_2) + [1 - p_0(\Delta t)]\varphi^{1/m}(u_1)\varphi^{1/m}(u_2) \\ &= 1 + m^{-1}\{p_0(\Delta t) \ln \varphi(u_1 + u_2) \\ &\quad + [1 - p_0(\Delta t)] \ln[\varphi(u_1)\,\varphi(u_2)]\} + o(m^{-1}).\end{aligned}$$

Consequently,

$$\begin{aligned}\varphi_{2m}^{m}(T_2; u_1, u_2) &\to [\varphi(u_1 + u_2)^{p_0(\Delta t)}\, [\varphi(u_1)\varphi(u_2)]^{1-p_0(\Delta t)} \\ &= \psi_2(T_2; u_1, u_2).\end{aligned}$$

It is easy to see that $\psi_2(T_2; u_1, u_2)$ is a characteristic function. In addition,

$$\psi_2(T_2; t, 0) = \psi_2(T_2; 0, t) = \varphi(t),$$

i.e., the consistency conditions are satisfied for $k = 2$. For $k > 2$ the proof is analogous. □

Note that for the Poisson flux with a parameter λ the probabilities p_s are

$$\begin{aligned}p_s = \exp&\left\{ -\lambda \sum_{j=1}^{n_s} (t_{q(s,j+1)-1} - t_{q(s,j)}) \right\} \\ &\times \prod_{j=1}^{n_s} \left[1 - \exp\{-\lambda(t_{q(s,j)} - t_{q(s,j)-1})\} \right].\end{aligned}$$

Lemma 6.3. If $|p_0''(t)| < c < \infty$ on $[0, T]$ and $\xi \geq 0$, then

$$\mathbf{M}\{\zeta_m(t_3) - \zeta_m(t_2)|\,|\zeta_m(t_2) - \zeta_m(t_1)|\} \leq H(t_3 - t_1)^2$$

for $t_1 \leq t_2 \leq t_3$.

Proof. From (6.20) we get

$$|\zeta_m(t'') - \zeta_m(t')| \leq \sum_{i=1}^{\nu(\Delta t)} (\xi_i^{(m)\prime\prime} + \xi_i^{(m)\prime}),$$

where $\Delta t = |t'' - t'|$, $\nu(\Delta t)$ is the random number of the flux realizations corresponding to $\{\xi_i^{(m)}(t;k)\}$ which have at least one point in (t', t''). The quantities $\xi_i^{(m)\prime}$ and $\xi_i^{(m)\prime\prime}$ are independent and have the same distribution $\varphi_m(u) = \varphi^{1/2}(u)$. By the Wald identity [6.1] we get

$$\begin{aligned}\mathbf{M}\{\zeta_m(t_3) - \zeta_m(t_2)| \, |\zeta_m(t_2) - \zeta_m(t_1)|\} \\ &\le 4\mathbf{M}\{\nu(t_3 - t_2)\,\nu(t_2 - t_1)\}[\mathbf{M}\xi_i^{(m)}]^2 \\ &\quad + m\mathbf{P}\{[\xi_i^{(m)}(t_3) - \xi_i^{(m)}(t_2)][\xi_i^{(m)}(t_2) - \xi_i^{(m)}(t_1)] > 0\}\,\mathbf{M}\xi_i^{(m)2} \\ &\le H_0(t_3 - t_2)(t_2 - t_1) \le H(t_3 - t_1)^2;\end{aligned}$$

hence,

$$\begin{aligned}\mathbf{P}\{[\xi_i^{(m)}(t_3) - \xi_i^{(m)}(t_2)[\xi_i^{(m)}(t_2) - \xi_i^{(m)}(t_1)] > 0\} \\ \le c_2(t_3 - t_2)(t_2 - t_1),\end{aligned}$$

$$\mathbf{M}\{\nu(t_3 - t_2)\,\nu(t_2 - t_1)\} \le c_1 m^2 (t_3 - t_2)(t_2 - t_1),$$

$$\mathbf{M}\xi_i^{(m)} = \mathbf{M}\xi/m, \quad \mathbf{D}\xi_i^{(m)} = \mathbf{D}\xi/m. \qquad \square$$

Lemma 6.3 shows that the Chentsov-Kolmogorov criterion is satisfied (Sect. 1.2).

Theorem 6.1. Under the conditions of Lemma 6.3 the processes $\zeta_m(t)$ weakly converge to a process which is determined by the limit finite dimensional distributions of Lemma 6.2 with a correlation function $k(t)$ and with the one-dimensional distribution function $F_\xi(x)$.

The Gaussian distribution is not described by this theorem. However, in this case the convergence of $\zeta_m(t)$ can be easily derived from the following generalized criterion:

$$\begin{aligned}\mathbf{P}\{|\zeta_m(t_3) - \zeta_m(t_2)| \ge \lambda, \ |\zeta_m(t_2) - \zeta_m(t_1)| \ge \lambda\} \\ \le \lambda^{-2\gamma}(t_3 - t_1)^{2\alpha}, \quad \gamma = \alpha = 1.\end{aligned}$$

All arguments considered up to Lemma 6.2 can be generalized to the case of random fields where obvious changes in the expressions for ψ_k, $k > 2$ must be made.

Let us now study the continuity of the limit process $\zeta(t)$ defined in Theorem 6.1. The most convenient criterion for the process $\zeta(t)$ to be continuous with a probability one is the Kolmogorov inequality [6.5])

$$m(p,h) = \mathbf{M}|\zeta(t+h) - \zeta(t)|^p \le ch^{1+r}, \qquad (6.21)$$

where $p > 0$, $r > 0$. Thus, the problem is to estimate the quantity $m(p,h)$. Note that by Lemma 6.2 the characteristic function of the vector $\{\zeta(t), \zeta(t+h)\}$ has the form

$$\varphi_h(u_1, u_2) = [\varphi(u_1 + u_2)]^{p_0(h)} [\varphi(u_1)\,\varphi(u_2)]^{1-p_0(h)}, \tag{6.22}$$

where $\varphi(u) = \varphi_\xi(u)$ is the characteristic function of the one-dimensional distribution of the process.

We now show that a random vector with the characteristic function (6.22) can be represented in a special simple form. We first note that $\varphi^\alpha(u)$ for arbitrary $\alpha > 0$ is a characteristic function. This follows from three points: (1) α can be represented as a limit of a sequence of rational numbers of the form m/n; (2) $\varphi^{m/n}(u)$ is a characteristic function, because the distribution of ξ is infinitely divisible; (3) the limit of a sequence of characteristic functions is also a characteristic function.

Thus, both factors in the right-hand side of (6.22) are characteristic functions of random vectors. The first vector has components which are all equal to probability one while the components of the second vector $\{\eta(t),\ \eta(t+h)\}$ are independent and distributed with the characteristic function $[\varphi(u)]^{1-p_0(h)}$. Therefore,

$$m(p, h) = \mathbf{M}\ |\eta(t+h) - \eta(t)|^p = \mathbf{M}|\mu_h|^p.$$

It is sufficient to estimate $m(p, h)$ for even values of p , since for a continuous process the quantities $m(p, h)$ are decreasing as $p \to \infty$.

The characteristic function of the random quantity μ_h has the form

$$\varphi_1(u) = [\varphi(u)\,\varphi(-u)]^{1-p_0(h)} \quad ;$$

therefore, the semi-invariants of the distribution of μ_h have the order (as $h \to 0$)

$$1 - p_0(h) = O(h).$$

Using the known relation between the semi-invariants and the central statistical moments [6.1], we get

$$m(p, h) = \mathbf{M}\ |\zeta(t+h) - \zeta(t)|^p = O(h)$$

for even p. Consequently, (6.21) is not applicable here for verifying the continuity of the limit process $\zeta(t)$. However the process described is continuous.

For example, the following criterion for a Gaussian process to be continuous with probability one is known [6.5]:

$$\mathbf{D}\{\zeta(t+h) - \zeta(t)\} \le K\,|\ln|h|\,|^{-p}, \quad p > 3.$$

This criterion is satisfied; hence, in the case of Gaussian distribution function, (6.20) is weakly convergent to a continuous Gaussian process. Perhaps there exist other distributions (e.g., for the Gamma-distributions with a large parameter) for which the limit process is continuous (or almost continuous). This assumption was confirmed by numerical simulations [6.6].

It is clear that this theorem can be generalized to random fields using the generalized Chentsov-Kolmogorov criterion. The finite-dimensional distributions of the random fields models described can be improved by using sums as in (6.20). Thus, by using the stratified isotropic models, it is possible to construct random fields with realistic realizations. Note that when random models of type (6.20) are used in the radiative transfer theory, it is sufficient to take m such that the characteristic size of the domain where $\zeta_m(r)$ is constant is less than the free path length of a photon.

We also mention the following useful property of the models proposed: it is possible to use a dependent simulation [on the basis of one Palm sequence corresponding to a given function $k(t)$] of fields with different one-dimensional distributions. Numerical simulation of these models shows that non-Gaussian fields have regions of constant values such that their diameters are essentially larger than the correlation scale of the field.

This fact can be interpreted as a manifestation of the intropy maximality of the Gaussian distribution when co-variances are fixed.

6.4 Simulation of Homogeneous Gaussian Fields by Randomization of the Spectral Representation

Let $\xi(x)$ be a real valued random Gaussian field with a correlation function $k(x)$ having a given spectral expansion

$$k(x) = \int \cos(\lambda\, x) p(\lambda)\, d\lambda \quad , \tag{6.23}$$

where $p(\lambda)$ is the spectral density, and $\lambda \in \Lambda = R_n$. Note that (6.23) exists , for example, if

$$\int |k(x)|\, dx < \infty.$$

Assume for simplicity that

$$\mathbf{M}\xi(x) = 0, \quad \mathbf{D}\xi(x) = 1.$$

Methods of simulation of Gaussian random processes and fields are well-developed for rational spectral densities. They are based on linear transformation of the white noise with appropriate spectral characteristics. The cost of these methods is sometimes too high, since they require a numerical solution of differential equations (or partial differential equations in the case of random fields). Next we consider a very simple, although not perfectly adequate, method of approximate construction of a Gaussian random field with a given correlation function as in (6.23).

Divide the space Λ into m parts, $\Lambda_1, \ldots, \Lambda_m$ and assume that the random points $\lambda_1, \ldots, \lambda_m$ are distrtibuted according to the probability densities

$$p_k(\lambda) = p(\lambda) \left[\int_{\Lambda_k} p(\lambda)\, d\lambda \right]^{-1}, \quad \lambda \in \Lambda_k.$$

Then by (6.23), we have

$$k(x) = \mathbf{M} \sum_{k=1}^{m} p_k \cos(\lambda_k x), \quad p_k = \int_{\Lambda_k} p(x)\, dx.$$

We thus obtain the following method for constructing a random field with a given correlation function:

(1) Random values $\lambda_1, \ldots, \lambda_m$ are sampled.

(2) Realization of the field is constructed by

$$\zeta_m(x) = \sum_{k=1}^{m} p_k^{1/2} [\xi_k \sin(\lambda_k x) + \eta_k \cos(\lambda_k)] \quad , \tag{6.24}$$

where $\{\xi_k, \eta_k\}$ is a set of independent standard Gaussian random values.

Note that ξ_1 [for $m = 1$ in 6.24)] presents a well-known model of the homogeneous random field [6.4].

One-dimensional conditional distribution of (6.24) is normal, i.e., Gaussian (and standardized) provided that $\lambda_1, \ldots, \lambda_m$ are fixed. Consequently, absolute one-dimensional distribution of (6.24) is also normal and standardized.

Similar arguments show that the multi-dimensional distributions of (6.24) are not Gaussian and the field is not ergodic. However, these disadvantages may relax if the space λ is divided into uniformly small parts as $m \to \infty$.

Convergence of processes and fields of form (6.24) as $m \to \infty$ is considered below. For the sake of convenience, we rewrite (6.23) in a general complex form

$$k(x) = \int \exp(i\lambda x) p(\lambda)\, d\lambda. \tag{6.25}$$

The corresponding complex normal field with a correlation function (6.25) can be defined by the spectral distribution

$$\zeta(x) = \int \exp(i\lambda x) \Phi(d\lambda) \quad , \tag{6.26}$$

where $\Phi(d\lambda)$ is the complex stochastic measure such that the random value $\int_{\Lambda} \Phi(d\lambda)$ is normal for every measurable domain $\Lambda_k \in \Lambda$, where

$$\mathbf{M} \int_{\Lambda_k} \Phi(d\lambda) = 0, \quad \mathbf{D} \int_{\Lambda_k} \Phi(d\lambda) = \int_{\Lambda_k} p(\lambda)\, d\lambda = p_k,$$

and

$$\mathbf{M} \left[\int_{\Lambda_k} \Phi(d\lambda) \int_{\Lambda_j} \Phi(d\lambda) \right] = 0$$

if the domains Λ_k and Λ_j are nonoverlapping.

In the complex case, (6.24) can be rewritten as follows:

$$\zeta_m(x) = \sum_{k=1}^{m} \exp\{i\lambda_k x\} \int_{\Lambda_k} \Phi(d\lambda). \tag{6.27}$$

Representation (6.27) gives (6.24) if the imaginary part of Φ as a function of λ is odd and the population $\{\Lambda_k, \lambda_k\}$ is invariant under sign changing.

Now consider the convergence of $\zeta_m(x)$ to a normal field with a given spectral density $p(\lambda)$. As mentioned above, this convergence is connected with the partitioning of Λ into subspaces Λ_k. Let V_k be the value of Λ_k, and let d_k be the diameter of Λ_k. We shall use uniform (with respect to the volume) partitions of the domain $\Lambda \backslash \Lambda_k$ such that

$$d_k \le c V_k^{1/n}, \quad k = 1, \ldots, m-1 \quad ,$$

where $V_k \to 0$ as $m \to \infty$. Such a partition can be obtained, for example, by using a uniform rectangular mesh with a step size $m^{-1/n}$.

Theorem 6.2. Let $\zeta(x)$ and $\zeta_m(x)$ be determined by (6.26, 27), respectively, for $|x| < R$, $|\lambda_m = \{\lambda : |\lambda| > l_m\}$ and $d_k < c l_m m^{-1/n}$, $k = 1, 2, \ldots, m-1$. Then (1) if $l_m m^{-1/n} \to 0$ and $l_m \to \infty$ as $m \to \infty$, then

$$\mathbf{M} \int_{|x|<R} [\zeta_m(x) - \zeta(x)]^2 \, dx \to 0;$$

(2) if for some $\varepsilon > 0$

$$\int |\lambda|^{\varepsilon} p(\lambda) \, d\lambda < \infty,$$

then for $l_m = c m^{2/[n(2+\varepsilon)]}$ the following estimation holds:

$$\mathbf{M} \int_{|x|<R} [\zeta_m(x) - \zeta(x)]^2 \, dx \le c m^{-2\varepsilon/[n(2+\varepsilon)]}.$$

Proof. It follows from (6.26, 27) that

$$
\begin{aligned}
&\mathbf{M} \int_{|x|<R} [\zeta_m(x) - \zeta(x)]^2 \, dx \\
&= \mathbf{M} \int_{|x|<R} dx \left\{ \sum_{k,j=1}^{m} \left[\int_{\Lambda_k} [\exp\{i\lambda_k x\} - \exp\{i\lambda x\}] \right. \right. \\
&\qquad \left. \left. \times \Phi(d\lambda) \int_{\Lambda_j} [\exp\{i\lambda_j x\} - \exp\{i\lambda x\}] \, \Phi(d\lambda) \right] \right\} \\
&= \int_{|x|<R} dx \sum_{k=1}^{m} \mathbf{M} \left[\int_{\Lambda_k} [\exp\{i\lambda_k x\} - \exp\{i\lambda x\}] \Phi(d\lambda) \right]^2 \\
&= \int_{|x|<R} dx \sum_{k=1}^{m} \int_{\Lambda_k} |\exp\{i\lambda_k x\} - \exp\{i\lambda x\}|^2 p(\lambda) d\lambda \\
&\le c_2 l_m^2 m^{-2/n} \int_{|\lambda|<l_m} p(\lambda) d\lambda + c_3 \int_{|\lambda|>l_m} p(\lambda) d\lambda .
\end{aligned} \tag{6.28}
$$

The first statement is thus proved. The second statement of the theorem follows from

$$
\int_{|\lambda|>l_m} p(\lambda) d\lambda \le l_m^{-\varepsilon} \int_{|\lambda|>l_m} |\lambda|^{\varepsilon} p(\lambda) d\lambda . \qquad \square
$$

Note that the estimate is optimal (by order of magnitude) if l_m is determined from

$$
P_m = \int_{|\lambda|>l_m} p(\lambda) d\lambda = l_m^2 \, m^{-2/n} .
$$

In addition, if $p(\lambda)$ decreases rapidly enough (e.g., exponentially) as $|\lambda| \to \infty$, then P_m is bounded by the quantity $c^* m^{-2/n} L(m)$, where $L(m)$ is a slowly increasing function. The convergence (in probability) of the L_2-error $\varepsilon = \|\zeta_n(x) - \zeta(x)\|_{L_2}$ to zero, where $\zeta(x)$ is a normal random field with a given density, follows by the Chebyshev inequality. Similar statements can be formulated and proved for the convergence of the derivatives of $\zeta_n(x)$ under the condition that appropriate moments of the spectral density exist. For example,

$$
\mathbf{M} \int [\zeta_m^{(k)}(x) - \zeta^{(k)}(x)]^2 \, dx \le c m^{-2\varepsilon/[n(2+\varepsilon)]} \quad \text{if} \tag{6.29}
$$

$$
\int |\lambda|^{2+\varepsilon} \, p(\lambda) \, d\lambda < \infty \quad , \tag{6.30}
$$

where $\zeta^{(k)}$ is the partial derivative of ζ with respect to $\lambda^{(k)}$, $k = 1, 2, \dots, n$.

Note that sometimes it is possible to prove the convergence (in probability) to zero of uniform deviation of ζ_m from ζ using imbedding theorems. In particular, for random processes $\zeta = \zeta(t)$ (i.e., for $n = 1$) it follows from (6.29, 30) and by the imbedding theorem ($W_2^{(1)}$ in C) that

$$\mathbf{M} \sup_t |\zeta_m(t) - \zeta(t)| \le c_0 m^{-\varepsilon/[n(2+\varepsilon)]}.$$

Note that the estimates obtained can be used to investigate the convergence of the functionals studied, since convergence in probability implies weak convergence with respect to the appropriate metric [6.5].

6.5 Stochastic Problems of Radiative Transfer Theory

As a first example, let us consider the problem of radiative transfer in a medium represented as a random set of spherical inhomogeneities. It is assumed that the midpoints of the spheres form a spatial Poisson flux, i.e., the numbers of midpoints in nonoverlapping domains are independent and distributed according to the Poisson law. Intersections of the spheres are allowed. Denote by σ_1 and σ_2 the total cross sections (Sect. 1.3) of the medium inside and outside the spheres, respectively. Assume that $\max(\sigma_1, \sigma_2) = \sigma_1$. Simulation of trajectories for a fixed realization of such a medium can be carried out by the method of maximal cross section (Sect. 1.3). In this method, the free path length is sampled according to the formula $l = -\ln \alpha$, where α is a random variable uniformly distributed on $(0, 1)$. If the sampled collision point does not lie in a sphere, then a "delta-scattering" is simulated, i.e., the particle moves in the same direction with probability $(\sigma_1 - \sigma_2)/\sigma_1$. Sometimes it is useful to apply a weight technique. The variance of this method for the media studied can be estimated as described in Sect. 6.2.

To construct a realization of the medium, it is sufficient to sample N , the number of sphere midpoints according to the Poisson distribution and then to sample all the midpoints independently and uniformly in the entire domain [6.1]. To reduce the number of arithmetic operations, it is possible to divide the domain into parts and to sample the midpoints in one ortion only when the particle hits this part. Clearly, the size of the part must be larger than the radii of the spheres. The sampled portion of the point flux must be stored because the particle may repeatedly hit the corresponding part of the domain. Resampling of the portions of the flux is not in accordance with the randomization principle of Sect. 6.1, resulting in a bias of the estimate which can be neglected only for strong scattering anisotropy.

Another general model of a stochastic medium follows from the representation of the total cross section $\sigma(r)$ in the form

$$\sigma^{(m)}(r) = \sum_{i=1}^{m} \sigma_i^{(m)}(r), \tag{6.31}$$

where $\sigma_i^{(m)}$, $i = 1, \ldots, m$ are independent realizations of a homogeneous random field related to stationary point fluxes as described in Sect. 6.3. Recall that the correlation function of the field σ coincides with the correlation function of each term $\sigma_i^{(m)}$, while the one-dimensional distribution is determined by the composition rule. Therefore, (6.31) is suitable if the one-dimensional distribution of the field simulated is infinitely divisible.

Various properties of fields are considered in Sect. 6.3, where we define a class of correlation functions $k(r)$ for isotropic fields, which is close to the class of convex functions. This provides satisfactory models of random fields required for solving many practical problems of transfer theory. It should be noted that the only reliable information about the correlation characteristics of a realistic random field is often the correlation scale (e.g., spatial scale) ρ, which is usually defined by

$$\rho = \int_0^\infty k(t)\,dt.$$

However, ρ can be defined by

$$\rho_1 = \left[-\frac{k(0)}{2k''(0)}\right]^{1/2} \tag{6.32}$$

as well.

Note that $\rho = \infty$ in the case of the isotropic layer field discussed in Sect. 6.3. Therefore, the correlation scale of this model is determined by (6.32). The correlation scale ρ for more complicated parallelepipedal models, where the independent point fluxes are sampled on coordinate axes (Sect. 6.3) , is finite: $\rho < \infty$.

As follows from the arguments of Sect. 6.3, it is useful to take a large value of m in (6.31) since the finite dimensional distributions of the field $\sigma^{(m)}$ become absolute-continuous as $m \to \infty$. This property is natural for fields with absolute-continuous one-dimensional distributions. However, in the problems of transfer theory it is possible to choose m such that the mean size of regions where $\sigma^{(m)}$ is constant is essentially less than the mean free path of the particle. It is preferable to obtain such a value of m in preliminary calculations. Various algorithms simulating the free path of the particle in a medium with a total cross section of form (6.31) are considered in [6.6]. Perhaps the simplest but not the most effective is here an algorithm where the free path length is sampled independently for each term $\sigma_i^{(m)}(r)$ and then the minimum of the sampled length is chosen. The distribution of this quantity coincides with the physical distribution of the free path length.

Some results of Monte Carlo calculations of transfer problems for stochastic plane-parallel media are given in [6.6]. It was shown that these results agree with the asymptotics of the mean intensity of radiation propagating in such media [6.2].

These asymptotics show that it is important to take into account the stochastic inhomogeneity of the actual media. The asymptotics can also be used to check the Monte Carlo calculations. In certain cases, a coefficient in the asymptotics can be effectively calculated by the Monte Carlo method. Therefore, we will consider these asymptotics in detail. It is well-known [6.8] that the asymptotics of the radiation intensity in a plane-parallel layer are determined to within a constant factor by solution of the transfer equation in an infinite medium. Correspondingly, the following model was introduced in [6.2].

(1) The medium density is defined as a homogeneous random field with plane symmetry; more exactly, the scattering and absorption coefficient depend on a single coordinate z, namely,

$$\sigma_s = \sigma_s(z) = q\sigma(z),\ \sigma_c = \sigma_c(z) = (1-q)\sigma(z),\ 0 \le q < 1$$

and the scattering indicatrix is the same everywhere. The random function $\sigma(z)$ is homogeneous (with respect to z) and satisfies the conditions of the Central Limit Theorem [6.1] for

$$\tau(z) = \int_0^z \sigma(z')\,dz'.$$

(2) The source is uniformly distributed on the plane $z = 0$, with the angular distribution of the source intensity satifying the chracteristic equation of the transfer theory [6.8]. Therefore , the radiation intensity for a given realization of σ is fully determined by the optical length measured from the observation point to the plane $z = 0$:

$$I(z;\sigma) = I(\tau(z)) = \exp(-\tau/L), \quad \tau(z) = \int_0^z \sigma(z')\,dz',$$

where L is the diffusion length (L equals the first eigenvalue of the characteristic equation for $\sigma = 1$, $\sigma_s = q$, i.e., it has a dimensionless diffusion length for the given scattering indicatrix and the fixed survival probability of a quantum in the scattering act).

(3) It is necessary to obtain the asymptotics of the mean intensity as $z \to \infty$, i.e., to find the expectation $\mathbf{E}\,I(z,\sigma)$. The formulated problem is thus generalized for an arbitrary statistical moment of the random intensity $\mathbf{E}\,I^k(z,\sigma)$ by replacing L with L/k.

The function $\exp(-x)$ is convex, thus we have by the Jensen inequality

$$\mathbf{E}I(z;\sigma) = \mathbf{E}\exp(-\tau/L) \ge \exp(-\mathbf{E}\tau/L) = \exp(-\sigma_0\, z/L), \quad \sigma_0 = \mathbf{E}\sigma.$$

Consequently, if

$$\mathbf{E}I(z;\sigma) \approx c\exp(-\alpha z), \quad z \to \infty,$$

then $\alpha \le \sigma_0/L$. Using the Central Limit Theorem for homogeneous random functions, the author obtained [6.2] the following asymptotics:

$$\mathbf{E}I(z;\sigma) \approx \exp\left\{-\frac{z\sigma_0}{L}\left(1-\frac{s^2\rho}{L\sigma_0}\right)\right\}, \tag{6.33}$$

where $s^2 = \mathbf{D}\sigma$ and ρ is the correlation length of the random function $\sigma(z)$. The asymptotics (6.33) are applicable , as mentioned in [6.2], if

$$2s^2\rho < L\sigma_0.$$

Formula (6.33) is not applicable for evaluating the high moments of the random intensity. Consequently, it is reasonable to try to estimate the desired asymptotics under some general assumptions for a physically justifable model of distribution of τ or for some simple model of $\sigma(z)$. We now obtain the asymptotics of the mean intensity for an "almost Gaussian" model for the distribution of the optical depth.

The Central Limit Theorem mentioned above implies that the distribution of $\tau(z)$ is asymptotically normal with parameters $\sigma_0 z$, $2z\rho s^2$ (in short we say that $\tau(z)$ is asymptotically $N(\sigma_0 z,\, 2z\rho s^2)$. We now assume that the distribution of τ is obtained as follows: the tail of the negative values of the distribution $N(\sigma_0 z,\, 2z\rho s^2)$ is concentrated at a point at $\tau = 0$, i.e., for $\eta = (\tau - \sigma_0 z)/(2z\rho s^2)^{1/2}$ we have

$$f_\eta(t,z) = \frac{1}{\sqrt{2\pi}}\exp(t^2/2) + \delta\left(t + \frac{\sigma_0}{s}\sqrt{\frac{z}{2\rho}}\right) F\left(-\frac{\sigma_0}{s}\sqrt{\frac{z}{2\rho}}\right), \tag{6.34}$$

for $t \geq -\frac{\sigma_0}{s}\sqrt{\frac{z}{2\rho}}$. Here $F(\cdot)$ is the distribution function of a standard Gaussian random variable. The distribution with the density (6.34) is called "almost Gaussian".

Substituting (6.34) into the integral representing the mean intensity $\mathbf{E}I(z)$ yields

$$\mathbf{E}I(z) = \exp\left\{-\frac{z\sigma_0}{L}\left(1-\frac{s^2\rho}{L\sigma_0}\right)\right\} + A(z).$$

Here

$$A(z) = F\left(-\frac{\sigma_0}{s}\sqrt{\frac{z}{2\rho}}\right) - \exp\left\{-\frac{z\sigma_0}{L}\left(1-\frac{s^2\rho}{L\sigma_0}\right)\right\}$$
$$\times\, \Phi\left(-\frac{\sigma_0}{s}\sqrt{\frac{z}{2\rho}}; -\frac{s\sqrt{2z\rho}}{L}\right),$$

where $\Phi(\cdot; M)$ is the distribution function $N(M,1)$.

The asymptotic analysis of $A(z)$ carried out for two cases: $2s^2\rho < L\sigma_0$ and $2s^2\rho > L\sigma_0$ on the basis of the asymptotics of the tail of the Gaussian distribution

$$\frac{1}{\sqrt{2\pi}}\int_{-\infty}^{-x} \exp(-t^2/2)\,dt \approx \frac{1}{\sqrt{2\pi x}}\exp(-t^2/2), \quad x \to \infty$$

gives the desired asymptotics for the "almost Gaussian" distribution of the optical depth:

$$\mathbf{E}I(z) \approx \begin{cases} \exp\left\{-\frac{z\sigma_0}{L}\left(1-\frac{s^2\rho}{L\sigma_0}\right)\right\}, & 2s^2\rho < L\sigma_0, \\ \frac{2\rho s^3\sqrt{2\rho}}{\sigma_0(2\rho s^2 - L\sigma_0)\sqrt{2\pi z}} \exp\left\{-\frac{z\sigma_0}{L}\frac{L\sigma_0}{4s^2\rho}\right\}, & 2s^2\rho > L\sigma_0, \end{cases} \quad (6.35)$$

Note that the factor $1 - \frac{s^2\rho}{L\sigma_0}$ is greater than 0.5, while $\frac{L\sigma_0}{4s^2\rho} < 0.5$. It is also interesting to note that in the derivation of (6.35) we used the asymptotics of two tails: the left tail of the distribution $N(0,1)$ and the right tail of the distribution $N(-s\sqrt{2z\rho/L}, 1)$. These asymptotics are equal to within a constant factor, i.e., asymptotically, the mean intensity $\mathbf{E}I^{(0)}(z)$ of the radiation through zero optical depth has the same order of magnitude as the total mean intensity $\mathbf{E}I(z)$. It is not difficult to verify that

$$\mathbf{E}I^{(0)}(z)/\mathbf{E}I(z) \sim 1 - L\sigma_0/(2\rho s^2).$$

Expression (6.35) shows that the effective diffusion length L' in the stochastic medium is larger than that in the deterministic medium:

$$L' = \begin{cases} L[1 - s^2\rho/(L\sigma_0)]^{-1}, & 2s^2\rho < L\sigma_0, \\ L[\sigma_0 L/(4s^2\rho)]^{-1}, & 2s^2\rho > L\sigma_0. \end{cases}$$

Thus, the stochasticity of the medium can essentially intensify the transmission of the radiation.

6.6 A Stochastic Elasticity Problem

Let us consider the equation which describes the bending of a thin elastic plate $G = \{x_1, x_2 : 0 \le x_1, x_2 \le 1\}$:

$$\Delta\Delta\, u = \sigma(x_1, x_2), \quad u|_{\partial G} = \Delta u|_{\delta} G = 0.$$

We suppose that the load σ is a homogeneous random field with a spectral density $p(\lambda)$, $(\lambda \in R^2)$. It is necessary to calculate the probability characteristics of the solution $u(x, y)$, e.g., the mean solution and the correlation function.

The stated problem is usually usually by the deterministic methods as follows: for a set of samples of σ the problem is solved by a method accepted, e.g., by finite-difference methods and then, using the approximate samples u, the desired characteristics are calculated.

It is clear that the randomized Monte Carlo methods may have advantages, since for a given sample of σ it is necessary to construct unbiased estimates (if the covariance moments are to be evaluated, then two conditionally independent samples are used) for the solution especially if it is necessary to evaluate the mean solution at one (or several) point.

In [6.9] numerical calculations were carried out for the following case $p(\lambda) = c \exp(-\alpha\lambda)$. The spectral model (Sect. 6.4) for σ, for $m = 1$ was used. The correlation function

$$B_r = \mathbf{E}\{u(x)\, u(x^{(0)})\}, \quad r = |x - x^{(0)}|$$

was calculated for the points $x = (1/2+r, 1/2)$, $x^{(0)} = (1/2, 1/2)$ by simulating two conditionally independent trajectories of the ε-spherical process (Sect. 1.6) starting at $x^{(0)}$ and x. The method considering the right-hand side of the equation described in Sect. 1.6 was used. The fundamental solution of the equation $\Delta\Delta u = \delta(P - P_j)$ is given by

$$V(r) = (8\pi)^{-1}\, r^2 \ln r, \quad r = |P - P_j|;$$

therefore,

$$W(P, P_j) = \mathbf{M}\left\{\frac{|P - P_j|^2}{8\pi} \ln|P - P_j| + \sum_{k=1}^{N} \frac{d^2(\xi_{k-1})}{8\pi}[1 + \ln(|P - P_j|)]\right\}, \tag{6.36}$$

where $\{\xi_{k-1}\}$, $(k = 1, 2, \ldots)$ is the trajectory of the ε-spherical process starting from P which is sampled uniformly in G, i.e., $p(x) = 1$.

In Table 6.1 below we compare the Monte Carlo calculations with the method of alternating directions with Chebyshev parameters which was applied to two Poisson equations obtained by splitting of the biharmonic equation.

Table 6.1

$r = \lvert x - x^0 \rvert$	Double randomization	Finite-difference method
0.0	0.808	0.812
0.2	0.703	0.725
0.4	0.654	0.669
0.5	0.433	0.491

6.7 Simulation of Admixture Diffusion in Stochastic Velocity Fields

The Lagrangian description is known to be often more convenient for modelling turbulent diffusion than the Euclidian description of this phenomenon. In the Lagrangian model many particle trajectories are generated beginning from a moment $t = t_0$. However, this method for solving diffusion problems may fail because it is difficult to construct incompressible Gaussian vector random of fields with a given complicated spectrum. Direct realization of the spectral representation of such a field (Sect. 6.4) is possible only if the spectrum contains a relatively small number of frequencies. As mentioned, the methods based on linear transformations of the "white noise" [6.4] are not general purpose methods and involve the solution of partial differential equations. Therefore, simple methods of numerical construction of homogeneous fields with a given correlation function and a one-dimensional Gaussian distribution of the type derived in Sect. 6.4 are important. However, to solve a series of practical problems of turbulent diffusion, it is sufficient to incorporate only these statistical characteristics of the velocity field. Indeed, calculations carried out in [6.10] confirm this conclusion. Note that to check the calculations of the particle concentration, it is possible to apply successive correction of the field simulation using the modelling technique of Sect. 6.4. Effective algorithms of field simulation reduce the cost of the numerical solution of the problems of the turbulent diffusion. Therefore, the real strength of the Lagrangian Monte Carlo approach lies in its potential to handle more general circumstances and to calculate complicated characteristics of the turbulent diffusion.

We now present a method for calculating of the correlation function of the concentration field of particles from a distributed source. Special characteristics of the turbulent diffusion for which explicit representations are known only under restricted assumptions were calculated in [6.10].

Let $u(x,t)$ be a vector Gaussian velocity field, spatially homogeneous, isotropic and time stationary. This field is fully defined by the covariance tensor

$$B_{ij}(x,\tau) = \mathbf{E}[u_i(r+x,t+\tau)u_j(r,t)]$$

or by the corresponding spectral tensor

$$\Phi_{ij}(k,\tau) = \frac{1}{(2\pi)^3}\int B_{ij}(x,\tau)\exp\{-ikx\}\,dx.$$

It follows from the isotropy that $B_{ij} = \delta_{ij}B/3$, where δ_{ij} is the Kronecker symbol. Therefore, modelling of the vector field can be reduced to the simulation of three scalar fields. These fields can be constructed on the basis of (6.24), where ξ_k, η_k must be replaced by vectors $\xi_{\mathbf{k}}$, $\eta_{\mathbf{k}}$, respectively. The components of $\xi_{\mathbf{k}}$, $\eta_{\mathbf{k}}$ are independent standardized Gaussian random variables. The vector field obtained is incompressible and has a prescribed correlation function if its components are sampled in a plane perpendicular to the random vector λ_k and

$$\mathbf{M}|\xi_k|^2 = \mathbf{M}|\eta_k|^2 = 1.$$

Incompressibility means that

$$\frac{\partial u_1}{\partial x} + \frac{\partial u_2}{\partial y} + \frac{\partial u_3}{\partial z} \equiv 0.$$

The described simulation technique was applied in [6.10], where the numerical results were obtained using (6.24) with $D_1 = R^3$.

Let us consider a randomized estimate of the covariance of the concentration field. Assume that a passive pollutant from a source of capacity $q(r,t)$, normalized so that

$$\int\int q(r,t)\,dr\,dt = 1$$

is diffusing in a velocity field $u(r,t)$. The stochastic concentration field $s(r,t)$ satisfies the equation

$$\frac{\partial s}{\partial t} + u_1\frac{\partial s}{\partial x} + u_2\frac{\partial s}{\partial y} + u_3\frac{\partial s}{\partial z} = q. \tag{6.37}$$

Simulating the field u and solving (6.37) for all such realizations, it is not difficult to obtain statistical estimates of various characteristics of the concentration field. Namely, the expectation value $s_0(r,t) = \mathbf{E}\, s(r,t)$, the covariance function $R(r_1, r_2, t_1, t_2)$, etc. However, the cost of this method is too high because (6.37) must be solved many times. However, it is clear that in order to obtain a Monte Carlo estimate of the mean concentration s_0 it is sufficient to simulate a single random particle trajectory $\rho(t)$ for each field realization. A similar randomized estimate of the covariance $R(r_1, r_2, t_1, t_2)$ can be calculated by simulating two independent trajectories $\rho_1(t)$, $\rho_2(t)$ for each realization of $u(t)$. These trajectories start at points sampled at random according to the density $q(r,t)$.

This technique follows from

$$\mathbf{E}\{s(r_1,t_1)s(r_2,t_2)\} = p(r_1,r_2;t_1,t_2), \tag{6.38}$$

where $p(r_1,r_2;t_1,t_2)$ is the point distribution density of the random vectors $\rho_1(t)$, $\rho_2(t)$. The equality holds, because

$$\begin{aligned}
p(r_1,r_2;t_1,t_2) &= \int\int p(\rho_1,\rho_2;t_1,t_2)\,\delta(\rho_1 - r_1)\delta(\rho_2 - r_2)\,d\rho_1\,d\rho_2\\
&= \int\int \mathbf{E}[p(\rho_1,\rho_2;t_1,t_2|u)]\,\delta(\rho_1 - r_1)\delta(\rho_2 - r_2)\,d\rho_1\,d\rho_2\\
&= \mathbf{E}\int\int p(\rho_1,\rho_2;t_1,t_2|u)\,\delta(\rho_1 - r_1)\delta(\rho_2 - r_2)\,d\rho_1\,d\rho_2\\
&= \mathbf{E}[s(r_1,t_1)\,s(r_2,t_2)].
\end{aligned}$$

Simulating pairs of conditionally independent trajectories $\rho_1(t)$, $\rho_2(t)$ for u fixed, and using standard methods of mathematical statistics for estimating the multi-dimensional distribution density $p(r_1,r_2;t_1,t_2)$, it is thus possible to calculate the covariance

$$R(r_1,r_2,t_1,t_2) = \mathbf{E}\{s(r_1,t_1)s(r_2,t_2)\} - s_0(r_1,t_1)\,s_0(r_2,t_2).$$

Note that the averaging over an ensemble of random fields and over particle trajectories is carried out simultaneously. One usually constructs the mean integral estimates of an observed density, i.e., histograms are used. This method leads to the construction of estimates of covariances between the integrals of the random concentration over nonintersecting spatial domains D_k, $k = 1, 2, \ldots$. These estimates can be obtained using pairs of conditionally independent trajectories $\rho_1(t)$ and $\rho_2(t)$, because the following equality holds:

$$P(\rho_1(t_1) \in D_i \cap \rho_2(t_2) \in D_k) = \mathbf{E}\left[\int\limits_{D_i} s(r_1, t_1)\, dr_1 \int\limits_{D_k} s(r_2, t_2)\, dr_2\right] . \quad (6.39)$$

Equation (6.39) is proved in nearly the same way as (6.32), the difference being that the delta function is replaced by indicator functions of D_i, D_k. Numerical results obtained on the basis of (6.39) are presented in [6.10].

7. The Method of Multiple Splitting

The multiple splitting technique is constructed as a generalization of the method of single splitting described in Sect. 7.5. In this chapter we handle general questions of optimization of the splitting method and special numerical models and algorithms which sometimes permit a simple discovering of the optimal values of the parameters. We also study minimax and randomized splittiing methods.

7.1 Optimization of the Splitting Method

The method of multiple splitting is applied to calculations of functionals represented in a form of a multi-dimensional integral

$$I = \mathbf{M}\varphi(\xi) = \int \varphi(x_1, ..., x_{k+1})\mathbf{P}(dx_1...dx_{k+1}),$$

where the random vector $\xi = (\xi_1, ..., \xi_{k+1})$ is distributed according to a probability measure $\mathbf{P}(dx_1...dx_{k+1})$.

Let $n = (n_1, ..., n_k)$ be a vector whose coordinates are positive integers, and let

$$\eta_n = \frac{1}{n_1} \sum_{j_1=1}^{n_1} \cdots \frac{1}{n_k} \sum_{j_k=1}^{n_k} \varphi(\xi_1, \xi_2^{(j_1)}, ..., \xi_{k+1}^{(j_k)}), \tag{7.1}$$

where $\xi_i^{(1)}, ..., \xi_i^{(n_{i-1})}$ are independent and equally distributed according to the conditional probabilistic measure $\mathbf{P}(dx_i|x_1, ..., x_{i-1})$, for all $i = 2, ..., k+1$, random variables. The conditional measure is defined as $\mathbf{P}(dx_1...dx_{k+1})$ under the condition that $\xi_1 = x_1, ..., \xi_{i-1} = x_{i-1}$.

For $k = 1$ (7.1) coincides with the estimate of Sect. 1.5, i.e., for $k = 1$

$$\mathbf{M}\eta_n = I. \tag{7.2}$$

For arbitrary k this equality can be proved by induction on the basis of

$$\mathbf{M}\eta_n = \mathbf{M}\mathbf{M}(\eta_n|\xi_1),$$

where $\mathbf{M}(\eta_n|\xi_1)$ is the conditional expectation of η_n under the condition that ξ_1 is fixed. Note that $\mathbf{M}(\eta_n|\xi_1)$ is a function of ξ_1. The equality

$$\mathbf{D}\eta_n = D_1 + \frac{D_2}{n_1} + \frac{D_3}{n_1 n_2} + ... + \frac{D_{k+1}}{n_1 n_2 ... n_k} \tag{7.3}$$

can also be proved by induction. Here

$$D_1 = \mathbf{DM}(\eta_n|\xi_1),$$

$$D_{i+1} = \mathbf{MD}[\mathbf{M}(\eta_n|\xi_1, ..., \xi_{i+1})|\xi_1, ..., \xi_i], \quad i = 1, ..., k.$$

The mean number of operations required for calculating one sample of η_n is defined by

$$T\eta_n = T_1 + n_1 T_2 + n_1 n_2 T_3 + ... + n_1 ... n_k T_{k+1},$$

where

$$T_{i+1} = \mathbf{M}T_{i+1}(\xi_1, ..., \xi_i) \quad .$$

$T_{i+1}(\xi_1, ..., \xi_i)$ is the mean number of operations corresponding to one sample of ξ_{i+1} provided $\xi_1, ..., \xi_i$ are fixed.

The optimal value $n = n^{(0)}$ is obtained as the solution to the following problem:

$$\min_n S_n, \quad S_n = T\eta_n \mathbf{D}\eta_n, \tag{7.4}$$

where min is taken over all positive integers k. It is known [7.1, 2] that it is useful to use the following approximate algorithm for finding $n^{(0)}$: first, (7.4) is solved for real values of k, next the integer valued vector closest to this optimal real valued vector is taken. This procedure makes sense, since the splitting method is applied when the value of $n_i^{(0)}$ is large: $n_i^{(0)} \gtrsim 3 (i = 1, ..., k)$. We denote by $n^{(0)}$ the solution to (7.4) on the set of positive real n. In [7.2] the following relations are given:

$$n_i^{(0)} = \sqrt{\frac{D_{i+1}}{D_i} \frac{T_i}{T_{i+1}}}, \quad i = 1, ..., k, \tag{7.5}$$

$$S_{n^{(0)}} = (\sqrt{T_1 D_1} + ... + \sqrt{T_{k+1} D_{k+1}})^2. \tag{7.6}$$

These relations can be justified by induction. For $k = 1$ the proof is obvious (Sect. 1.5). Assume that $n_2, ... n_k$ are fixed. Then

$$\mathbf{D}\eta_n = D_1 + \frac{1}{n_1}\left(D_2 + \frac{D_3}{n_2} + ... + \frac{D_{k+1}}{n_2 ... n_k}\right) = D_1 + \frac{D_2^{(1)}}{n_1},$$

$$T\eta_n = T_1 + n_1(T_2 + n_2 T_3 + ... + n_2 ... n_k T_{k+1} = T_1 + n_1 T_2^{(1)}.$$

Minimizing the quantity S_n with respect to n_1 yields

$$n_1^{(1)} = \sqrt{\frac{D_2^{(1)}}{D_1} \frac{T_1}{T^{(1)}}}, \tag{7.7}$$

$$S_n^{(1)} = \left(\sqrt{T_1 D_1} + \ldots + \sqrt{T_2^{(1)} D_2^{(1)}}\right)^2 . \tag{7.8}$$

Consequently, to complete the proof, it is necessary to verify that

$$\frac{\mathbf{D}\eta_{n^{(0)}}}{T\eta_{n^{(0)}}} = \frac{D_1}{T_1}. \tag{7.9}$$

This relation is evident if $k = 1$. Note that in view of (7.8) the problem of minimization of S_n reduces to an analogous problem of minimization of $T_2^{(1)} D_2^{(1)}$ whose dimension is, however, less.

We now assume that the relations (7.5, 6, 9) are satisfied for $k-1$. Then from (7.7–9) it follows that these relations remain valid for k. This completes the justification of (7.5, 6). Denote the quantities $D\eta_n, T\eta_n$ for $n = (1, \ldots, 1)$ by D and T, respectively. Then the computational advantage due to the optimal splitting is defind by

$$\frac{TD}{S_{n^{(0)}}} = \frac{(T_1 + \ldots + T_{k+1})(D_1 + \ldots + D_{k+1})}{(\sqrt{T_1 D_1} + \ldots + \sqrt{T_{k+1} D_{k+1}})^2}.$$

7.2 Optimization of the Splitting Technique for Calculating the Transmission Probability

Suppose that it is required to calculate the probability p that a particle transmits through a layer of material. To construct the splitting method we divide the layer into $k+1$ sublayers according to the uniform probability. This means that the transmission probabilities for all the layers are equal to $p^{1/(k+1)}$. When intersecting the i-th surface, the trajectory is splitting in n_i $(i = 1, \ldots, k)$ parts. The value $i = 0$ corresponds to the surface of the source, and the value $i = k+1$ to the surface where the transmitted particles are recorded. It is natural to introduce a random vector $\xi = (\xi_1, \ldots, \xi_{k+1})$, such that $\xi_i = 1$ if the trajectory has intersected the i-th surface, and $\xi_i = 0$ otherwise. Now,

$$p = \mathbf{M}\varphi(\xi_1, \ldots, \xi_{k+1}),$$

where $\varphi \equiv 1$, if $\xi_{k+1} = 1$, and $\varphi \equiv 0$, if $\xi_{k+1} = 0$.

Thus, the trajectory splitting is formulated as a scheme described in Sect. 7.1. Simple arguments show that

$$D_i = p^{2(k+1-i)/(k+1)} p^{i/(k+1)} (1 - p^{1/(k+1)}), \quad i = 1, \ldots, k+1.$$

We assume that

$$T_{i+1} = T_i p^{1/k+1)},$$

i.e., the mean computer time for one trajectory from i-th to $(i+1)$-th splitting does not depend on i. Then

$$n_i^{(0)} = m = \sqrt{\frac{D_{i+1}}{D_i}\frac{T_i}{T_{i+1}}} = p^{-1/(k+1)}. \quad i = 1, ..., k$$

and

$$\mathbf{D}\eta_{n^{(0)}} = D_1(k+1), \quad T\eta_{n^{(0)}} = T_1(k+1).$$

The computational advantage due to the splitting scheme is here determined by

$$\frac{TD}{S_{n^{(0)}}} = \frac{(1-p)^2}{(k+1)^2 p^{k/(k+1)}(1-p^{1/(k+1)})^2}.$$

Note that if n is chosen optimally, then an addition of new splitting surfaces does not increase the computational cost. However, practical calculations show that k must be chosen so that $n_i^{(0)} \gtrsim 3 (i = 1, ..., k)$. Otherwise the round-off errors of these quantities make the efficiency of the estimate under study worse.

We now assume that the depth of the layer is increasing by means of an addition of new sublayers, i.e.,

$$p = p_0^{k+1}. \quad p_0 = \text{const}, \quad k \to \infty.$$

Under this condition it is not difficult to derive the following asymptotic relation

$$\frac{TD}{S_{n^{(0)}}} \sim \frac{1}{(1+k)^2 p}.$$

This yields

$$S_{n^{(0)}} \sim (1+k)^2 p^2 \quad ,$$

since $T \sim$ const and $D \sim p$.

If we take the mean number of operations needed to achieve the given relative accuracy as a measure of the computational cost, then

$$S_{n^{(0)}}^{(0)} \sim (1+k)^2.$$

This means that the relative cost of the optimal splitting algorithm is increasing as a square of the depth of the layer.

7.3 Numerical Calculation of the Optimal Splitting Parameters

It is difficult to calculate the quantities D_i and T_i by the direct Monte Carlo method, since additional codes must be written [7.2]. However, by calculating the quantities $\mathbf{D}\eta_n, T\eta_n$ for a series of values of n, it is then possible to construct a system of equations for the unknowns D_i and T_i:

$$D_1 + \frac{D_2}{n_1^{(t)}} + \ldots + \frac{D_{k+1}}{n_1^{(t)} \ldots n_k^{(t)}} = \tilde{D}\eta_{n^{(t)}},$$

$$T_1 + n_1^{(t)} T_2 + \ldots + n_1^{(t)} \ldots n_k^{(t)} T_{k+1} = \tilde{T}\eta_{n^{(t)}}, \quad t = 1, \ldots, k+1. \tag{7.10}$$

If the calculations of the vector n^* with uniformly maximal coordinates are carried out, then it is possible to calculate simultaneously the right-hand sides in (7.10). The positive correlation improves the results [7.2].

Sometimes, it is possible to construct the splitting method so that the chain of splittings is practically homogeneous, and

$$D_i / D_{i+1} = a, \quad T_i / T_{i+1} = b, \quad i = 1, \ldots, k. \tag{7.11}$$

The test mentioned in Sect. 7.2 possesses this property. The results reported in [7.2] agree with (7.11). If (7.11) is satisfied, then

$$n_1^{(0)} = n_k^{(0)} = m = \sqrt{b/a}.$$

Calculating $D\xi$ and t for two values, m_1, m_2, we can use the following equations for a and b:

$$\begin{aligned} \frac{\tilde{\mathbf{D}}\eta_{m_1}}{\tilde{\mathbf{D}}\eta_{m_2}} &= \left(1 + \sum_{i=1}^{k} \frac{a^i}{m_1^i}\right)\left(1 + \sum_{i=1}^{k} \frac{a^i}{m_2^i}\right)^{-1} = \Psi_1(a), \\ \frac{\tilde{T}\eta_{m_1}}{\tilde{T}\eta_{m_2}} &= \left(1 + \sum_{i=1}^{k} m_1^i b^i\right)\left(1 + \sum_{i=1}^{k} m_2^i b^i\right)^{-1} = \Psi_2(a). \end{aligned} \tag{7.12}$$

Simple arguments show that the functions $\Psi_1(a)$ and $\Psi_2(b)$ on $(0, \infty)$ are monotonic. Let us consider, for example, the function $\Psi_2(b)$ under the condition that $m_1 > m_2$. This function is monotonically increasing, if $k = 1$. We rewrite $\Psi_2(b)$ in the form

$$\Psi_2(b) = \Psi_2(b,k) = \frac{B_1(b,k)}{B_2(b,k)} = \frac{1 + bn_1 B_1(b,k-1)}{1 + bn_2 B_2(b,k-1)}.$$

This yields

$$\begin{aligned} B_2^2(b,k)\Psi_2'(b) = {} & b^2 n_1 n_2 B_2^2(b,k-1)\Psi_2'(b,k-1) \\ & + [bn_1 B_1(b,k-1)]' - [bn_2 B_2(b,k-1)]' \end{aligned}$$

which in turn proves by induction that the function $\Psi_2(b,k)$ is monotonically increasing. Analogously the monotonic decrease of the function $\Psi_1(a)$ under the condition $m_1 > m_2$ can be proved. Thus, the solutions to (7.12) are unique. We now prove that these solutions exist. Assume that $m_1 > m_2$. Then, obviously

$$m_2^i m_1^{-i} \le \Psi_1(a) \le 1, \quad a \ge 0,$$

$$1 \le \Psi_2(b) \le m_1^i m_2^{-i}, \quad b \ge 0.$$

As a result, it may appear that (7.12) are not resolvable.

This implies either the quantities $D\eta_n, T\eta_m$, were calculated with low accuracy or (7.11) is not applicable. For numerical purposes, it is convenient to rewrite (7.11) in the form

$$\begin{aligned} \frac{\tilde{D}\eta_{m_1}}{\tilde{D}\eta_{m_2}} &= \frac{1-a^{k+1}/m_1^{k+1}}{1-a/m_1}\,\frac{1-a/m_2}{1-a^{k+1}/m_2^{k+1}}, \\ \frac{\tilde{T}\eta_{m_1}}{\tilde{T}\eta_{m_2}} &= \frac{1-m_1^{k+1}b^{k+1}}{1-m_1 b}\,\frac{1-m_2 a}{1-m_2^{k+1}b^{k+1}}. \end{aligned} \tag{7.13}$$

As a first example, in [7.2] a problem of radiative transfer through a plane–parallel layer of a material with an isotropic scattering and with a surval probability $q = 0,9$ was solved. The probability of transmission p which was about $p = 0.0007$ was calculated by the splitting method. A six-fold splitting on planes uniformly dividing the layer was used. Calculations have given: $n^{(0)} = (2,3,3,3,3,3)$[7.2]. Note that in [7.2] (7.12) were used; the result was $m = 3$. Thus the equations (7.12) give a satisfactiry variant of splitting, although (7.12) is not exactly satisfied. It is interesting to note that the model of Sect. 7.2 gives $m = p^{-1/7} = 2.83$.

7.4 Uniform Optimization of the Splitting Method

We assume that the splittimg method can be applied to simultaneous calculation of a set of integrals of the type

$$I_j = \mathbf{M}\varphi_j(\xi) = \int \varphi_j(x_1,...,x_{k+1})\mathbf{P}(dx_1,...,dx_{k+1}),$$

$$j = 1,...,m.$$

Introduce a finite probabilistic measure $\Lambda = (\lambda_1,...,\lambda_m)$. Let $S_n(\Lambda)$ be a mean computational cost of the splitting method

$$S_n(\Lambda) = T^{(m)}\eta_n \mathbf{E}\mathbf{D}\eta_n,$$

where

$$\mathbf{E}\mathbf{D}\eta_n = \sum_{j=1}^{m} \lambda_j \mathbf{D}\eta_n^{(j)},$$

and $\eta_n^{(j)}$ is the Monte Carlo estimate of the splitting method for I_j and $T\eta_n$ is the mean computational time required for calculating the estimates $\eta_n^{(j)}$, i.e.,

$$T\eta_n^{(m)} = T_1(m) + n_1 T_2^{(m)} + ... + n_1...n_k T_{k+1}^{(m)}.$$

Note that sometimes the quantities $T_i^{(m)}$ weakly depend on m. The relation $T_i^{(m)} \approx T_i(1)(i = 1, ..., k+1)$ holds, e.g., in the transfer theory, if, to calculate $\{\varphi\}$, simple events like intersections with plates or collisions of different type are recorded.

Thus,

$$S_n(\Lambda) = \left(T_1^{(m)} + n_1 T_2^{(m)} + ... + n_1...n_k T_{k+1}^{(m)}\right) \times \left(\mathbf{E}D_1 + \frac{\mathbf{E}D_2}{n_1} + ... + \frac{\mathbf{E}D_k}{n_1...n_{k+1}}\right).$$

Note that in the derivation of (7.5, 6) we have not used any relations between D_i; thus, the expression for $\mathbf{E D}\eta_n$ has the form (7.3). Therefore,

$$n^{(0)} = \sqrt{\frac{\mathbf{E}D_{i+1}}{\mathbf{E}D_i}\frac{T_i^{(m)}}{T_{i+1}^{(m)}}} \quad \text{with} \quad i = 1, ..., k,$$

$$S_{n^{(0)}}(\Lambda) = \left(\sqrt{T_1^{(m)}\mathbf{E}D_1} + ... + \sqrt{T_{k+1}^{(m)}\mathbf{E}D_{k+1}}\right)^2.$$

Let us now consider the problem of uniform optimization of the splitting method which consists of finding

$$\min_n \max_j \mathbf{D}\eta_n^{(j)}. \tag{7.14}$$

From the known relation between the mean squared and minimax solutions (Sect. 4.1) we have the following lemma.

Lemma 7.1. The optimal value $n = n^*$ [as in (7.14)] is determined by

$$n_i^* = \frac{\mathbf{E}^* D_{i+1}}{\mathbf{E}^* D_i}\frac{T_i^{(m)}}{T_{i+1}^{(m)}}, \quad i = 1, ..., k,$$

where E^* corresponds to the measure Λ^* which minimizes the quantity $S_{n^{(0)}}(\Lambda)$.

Note, that in order to determine n_i^*, it is convenient to use the approach described in Sect. 7.3, which is based on equations as in (7.10, 13).

7.5 Randomized Splitting Method

We now assume that it is necessary to calculate

$$\begin{aligned} I &= \mathbf{E}I(\sigma) = \mathbf{EM}\varphi(\xi,\sigma) \\ &= \int \Lambda(d\sigma) \int \varphi(x_1, ..., x_{k+1}; \sigma)\mathbf{P}(dx_1, ..., dx_{k+1}). \end{aligned}$$

Here Λ is a probabilistic measure in a functional space of the parameter σ.

The randomized splitting method is based on

$$I = \mathbf{EM}\eta_n(\sigma). \tag{7.15}$$

Namely, for each sample of σ only one sample of $\eta_n(\sigma)$ is used.

Optimization of such an algorithm is obviously reduced to finding n which minimizes the quantity

$$\begin{aligned} S_n &= \mathbf{E}T\eta_n(\sigma)[\mathbf{EM}\eta_n^2(\sigma) - I^2] \\ &= \mathbf{E}T\eta_n(\sigma)[\mathbf{EM}\eta_n^2(\sigma) - \mathbf{D}I(\sigma)] \\ &= (\mathbf{E}T_1 + n_1\mathbf{E}T_2 + ... + n_1...n_k\mathbf{E}T_{k+1}) \\ &\quad \times \left(\mathbf{E}D_1 + \mathbf{D}I(\sigma) + \frac{\mathbf{E}D_2}{n_1} + ... + \frac{\mathbf{E}D_k}{n_1...n_{k+1}}\right). \end{aligned}$$

Consequently,

$$n_1^{(0)} = \sqrt{\frac{\mathbf{E}D_2}{\mathbf{E}D_1 + \mathbf{D}I(\sigma)}},$$

$$n_1^{(0)} = \sqrt{\frac{\mathbf{E}D_{i+1}}{\mathbf{E}D_i}\frac{\mathbf{E}T_i}{\mathbf{E}T_{i+1}}}, \quad i = 2, ..., k,$$

$$S_n = (\sqrt{\mathbf{E}T_1[\mathbf{E}D_1 + \mathbf{D}I(\sigma)]} + \sqrt{\mathbf{E}T_2\mathbf{E}D_2} + ... + \sqrt{\mathbf{E}T_{k+1}\mathbf{E}D_{k+1}})^2.$$

Note that the conditions of type (7.15) are usually satisfied in the transfer problems described in Sects. 7.2, 3. The method of Sect. 7.3 can be effectively applied to optimize the randomized splitting method.

7.6 Splitting of the Collision Estimate

Let us now study a possibility for reducing the computational cost of the Monte Carlo algorithm for solving integral equations by appropriate splitting [7.3]. The computational cost is understood to be the product $t\mathbf{D}\xi$, where t is the mean computational time for one sample, and $\mathbf{D}\xi$ is the variance. We assume that the cost can be represented as a sum of terms which correspond to different stages of the algorithm.

Using the formula of conditional expectation, it is possible to represent the variance in the form of an analogous sum.

The basic idea can be formulated as follows: the cost of a certain stage of the algorithm must be proportional to the corresponding contribution to the variance. In what follows we use the fact that for nonnegative a_i, b_i we have $\sum a_i \sum b_i \geq (\sum \sqrt{a_i b_i})^2$ which transforms to equality iff the ratio a_i/b_i does not depend on i.

Usually, the splitting of the trajectories is carried out when they intersect fixed planes; the relation between the splitting parameters and the method of simulation of the transitions is not used. We now give a more general splitting scheme. We will see that the mean number of new particles which appear at a given point might also depend on the previous state of the trajectory.

Assume that it is necessary to calculate the solution to

$$\varphi(x) = \int\limits_X k(x,x')\varphi(x')dx' + h(x), x \in X, \tag{7.16}$$

or $\varphi = K\varphi + h$, at a given point $x_0 \in X$. Here $X \subset \mathbf{R}_n, h \in L_\infty(X)$, and K is an integral operator in $L_\infty(X)$. We assume that the spectral radius of the operator K with the kernel $|k(x,x')|$ is less than 1.

Define a branching random process in X as follows. At the initial (zero) time instance there exists one particle at the point $x_0 = x$. Next x_0 passes to x_1 according to a transition density $r_1(x,x_1)$ and $q_1(x,x_1)$ new particles are generated on the average (with probability $\{q_1\}$ $[q_1]+1$ particles are generated, and $[q_1]$ particles are generated with probability $1 - \{q_1\}$). The braces $[\ \ ]$ and $\{\ \ \}$ denote the integer and fractional part, respectively. The quantity q_1 is called a survival probability.

If a particle of $(i-1)$-th generation is generated at the point x, then it passes, independently on other particles generated at x, to the point x' according to the distribution density $r'(x,x')$; at x' $q_i(x,x')$ new particles of i-th generation are generated on the average. It is assumed that the functions r_i, q_i may depend not only on x, x', but also on the whole of the history of the particle. On the tree of such a process it is possible to define an analog of the collision estimate ξ_x (Sect. 1.6). For this estimate the following recurrent relation corresponding to (7.16) holds:

$$\xi_x = h(x) + \frac{k(x,x_1)}{r_1(x,x_1)} \tag{7.17}$$
$$\times \left\{ h(x_1) + \frac{1}{q_1(x,x_1)} \left[\sum_{i=1}^{[q]} \left(\tilde{\xi}_{x_1}^{(i)} - h(x_1)\right) + \delta\left(\tilde{\xi}_{x_1}^{[q]+1} - h(x_1)\right)\right]\right\}.$$

Here $\tilde{\xi}^{(i)}$ are independent samples of the estimate $\tilde{\xi}$ which is also a collision estimate; $\delta = 0$ or $\delta = 1$, depending on the number of particles generated at the point x_1. The expression in the curly brackets has the expectation $\varphi(x_1)$ (x_1 fixed) and the variance

$$\frac{\mathbf{D}\tilde{\xi}_{x_1}}{q_1(x,x_1)} + \frac{\{q_1\}(1-\{q_1\})}{q_1^2(x,x_1)}[K\varphi]^2(x_1). \tag{7.18}$$

By the formula of conditional expectations we obtain from (7.18)

$$\begin{aligned}\mathbf{D}\xi_x &= D_1(x) + \int \frac{k^2(x,x_1)}{r_1(x,x_1)} \frac{\mathbf{D}\tilde{\xi}_{x_1}}{q_1(x,x_1)} dx_1 \\ &+ \int \frac{k^2(x,x_1)}{r_1(x,x_1)} \frac{\{q_1\}(1-\{q_1\})}{q_1^2(x,x_1)}[K\varphi]^2(x_1)dx_1,\end{aligned} \tag{7.19}$$

where $D_1(x)$ is the variance of the estimate for the integral

$$I(x) = \int k(x,x_1)\varphi(x_1)dx_1,$$

that is

$$D_1(x) = -[K\varphi]^2(x) + \int \frac{k^2(x,x_1)}{r_1(x,x_1)}\varphi^2(x_1)dx_1.$$

The average computational cost requried for obtaining one sample is represented in the form

$$t(x) = t_1(x) + \int r_1(x,x_1)q_1(x,x_1)\tilde{t}(x_1)dx, \tag{7.20}$$

where $t_1(x)$ is the average cost for one transition from x to x_1; $\tilde{t}(x_1)$ is the average cost for one sample of $\tilde{\xi}x_1$. For simplicity, we assume that the third term in (7.19) is small. Then we denote the sum of the two first terms in (7.19) by D_x and study the product $t(x)D_x$. The following inequalities are obvious:

$$\begin{aligned}t(x)D_x &= (t_1(x) + \int r_1(x,x_1)q_1(x,x_1)\tilde{t}(x_1)dx_1) \\ &\quad\times \left(D_1(x) + \int \frac{k^2(x,x_1)}{r_1(x,x_1)} \frac{\mathbf{D}\tilde{\xi}_{x_1}}{q_1(x,x_1)} dx_1\right) \\ &\geq \left(\sqrt{t_1(x)D_1(x)} + \left(\int r_1(x,x_1)q_1(x,x_1)t(x_1)dx_1\right.\right. \\ &\quad\left.\left.\times \int \frac{k^2(x,x_1)}{r_1(x,x_1)} \frac{\mathbf{D}\tilde{\xi}_{x_1}}{q_1(x,x_1)} dx_1\right)^{1/2}\right)^2 \\ &\geq \left(\sqrt{t_1(x)D_1(x)} + \int |k(x,x_1)|\sqrt{\tilde{t}(x)\mathbf{D}\tilde{\xi}_{x_1}}dx_1\right)^2.\end{aligned} \tag{7.21}$$

It is not difficult to verify that for arbitrary $\tilde{\xi}$ and $r_1(x,x_1)$ these inequalities turn into equalities if

$$q_1(x,x_1) = \sqrt{\frac{t_1(x)}{D_1(x)}} \frac{|k(x,x_1)|}{r_1(x,x_1)} \sqrt{\frac{\mathbf{D}\tilde{\xi}_{x_1}}{\tilde{t}(x_1)}}. \tag{7.22}$$

In addition, for such q_1 in this case the following equality holds:

$$\frac{D_x}{t(x)} = \frac{D_1(x)}{t_1(x)}. \tag{7.23}$$

In practice, only an approximate behaviour of the functions $t_1(x)/D_1(x)$, and $\mathbf{D}\,\tilde{\xi}_{x_1}/t(x_1)$ is known. This could be used to construct a practically optimal chain.

It can be seen from (7.21) that if $q_1(x, x_1)$ is taken according to (7.22), then the density r_1 must be found from the condition of minimization of the quantity $t_1(x)D_1(x)$. In addition, the ratio $t_1(x)/D_1(x)$ must be calculated as accurately as possible. This problem is difficult to formalate.

We assume that r_1 is chosen from some general arguments or preliminary calculations. Note that in many cases the densities r_1 used in practice are far from being optimal importance densities. For example, this is a case when the kernel $k(x, x')$ of (7.16) is complicated and there are many difficulties in calculating $[Kg](x)$ for an arbitrary function $g(x)$. Then large values of $D_1(x)/I(x)$ can be expected. Our results have significance exactly for such cases. The formula (7.21) shows that the estimate $\tilde{\xi}$ must be constructed so that $\tilde{t}D\tilde{\xi}$ will be minimal. Thus, all the previous arguments could be repeated for $\tilde{\xi}$. There is no reason to introduce a dependence of $\tilde{\xi}$ on x. We assume that the density $r_2(x, x')$ coincides with $r_1(x, x')$, because it is chosen from analogous arguments. Taking q_2 equal to q_1, from (7.22) we obtain a relation analogous to (7.23):

$$\frac{\tilde{D}_{x_1}}{\tilde{t}(x_1)} = \frac{\tilde{D}_1(x_1)}{\tilde{t}_1(x_1)} = \frac{D_1(x_1)}{t_1(x_1)}.$$

Now let

$$q_1(x, x_1) = \sqrt{\frac{t_1(x)}{D_1(x)} \frac{|k(x, x_1)|}{r_1(x, x_1)}} \sqrt{\frac{D}{\tilde{t}(x_1)}}. \tag{7.24}$$

For this q_1 we use the third term in (7.19):

$$\int \frac{k^2(x, x_1)}{r_1(x, x_1)} \frac{\{q_1\}(1 - \{q_1\})}{q_1^2(x, x_1)} [K\varphi]^2(x_1) dx_1$$
$$= \frac{D_1(x)}{t_1(x)} \int r_1(x, x_1)\{q_1\}(1 - \{q_1\}) \frac{[K\varphi]^2(x_1)}{D_1(x_1)} t_1(x_1) dx_1.$$

This expression is small compared to $D_1(x)$ if, for example, $t_1(x)$ weakly depends on x, and the density r_1 exists so that the relative variance of the estimate for the integral $I(x)$ is quite large [e.g., of order 10; the above expression is $2 - 3\%$ of $D_1(x)$]. On the contrary, if the density r_1 corresponds to the importance simulation technique, then this expression may play an essential role in (7.19).

The third term in (7.19) is small; thus, so we conclude that $\mathbf{D}\zeta_x \approx D_x$, and, analogously, $\tilde{\mathbf{D}}\xi_{x_1} \approx \tilde{D}_{x_1}$. Therefore, first, the survival coefficient (7.19) is close to the optimal value (7.22), and, second, the product $t(x)D_x$ which is minimized by the above choice of the survival coefficient is close to $t(x)\mathbf{D}\xi_x$.

Thus, the simulated process is most likely Markovian. If $r(x,x')$ is its transition density, and $a(x)$ approximately describes the behavior of $\sqrt{D_1(x)/t_1(x)}$, then the survival probability q_1 must be taken as

$$\frac{1}{a(x)}\frac{|k(x,x')|}{r(x,x')}a(x').$$

Note that in all of the above arguments we assumed that the quantities $\tilde{\mathbf{D}}\xi$ and $\tilde{t}$ are finite. Therefore, the finiteness of the variance and of the computational cost of such homogeneous estimate must be proved.

Let $\eta^{(m)}$ be the conditional expectation ξ under the condition that m first generations are fixed. Then

$$\mathbf{D}\eta_x^{(m)} = \sum_{k=0}^{m-1} a(x)\left[K_1^k\frac{D_1+\varepsilon}{a}\right](x),$$

where ε is the third term in (7.19). Note that $\eta^{(m)}$ tends in probability to ξ, since

$$\mathbf{M}|\xi-\eta^{(m)}| \le K_1^m|\varphi| + \sum_{n=m}^{\infty} K_1^m|h|,$$

where the right-hand side tends to zero as m increases. Consequently, a subsequence from $\eta^{(m)}$ can be extracted which tends to ξ with probability one. By the Lebesgue theorem we get

$$\mathbf{D}\xi_x = \lim \mathbf{D}\eta_x^{(m)} = a(x)\sum_{n=0}^{\infty}\left[K_1^n\frac{D_1+\varepsilon}{a}\right](x).$$

It is not difficult to verify that the mean number of points of the n-th generation is equal to $[K_1^n a](x)/a(x)$. Therefore, the mean number of transitions in the chain is equal to

$$\frac{1}{a(x)}\sum_{n=0}^{\infty}[K_1^n a](x).$$

8. Transformation of Equations and Weighted Estimates

In this chapter we give some transformations of the integral equations which permit to effective application of the Monte Carlo methods and improve the estimates in the sense of weak convergence.

8.1 The Averaging Transformation

Let us consider the integral equation in $L_\infty(X)$

$$\varphi(x) = \int_X k(x,x')\varphi(x')dx' + h(x), \tag{8.1}$$

or

$$\varphi = K\varphi + h.$$

Applying the operator K yields

$$K\varphi = K(K\varphi) + Kh. \tag{8.2}$$

Thus, the function $K\varphi$ also satisfies an equation of type (8.1) where h is replaced with

$$(Kh)(x) = \int_X k(x,x')h(x')dx'. \tag{8.3}$$

From (8.1) we have

$$K\varphi = \varphi - h.$$

Therefore, the estimate for the solution to (8.1) has the form

$$\xi_x^{(1)} = h(x) + \sum_{n=0}^{N} Q_n(Kh)(x_n). \tag{8.4}$$

On the basis of this estimate it is possible to construct the local estimate [8.1, 2] for the solution of the adjoint equation of the type

$$\varphi^*(x) = \int_X k(x', x)\varphi^*(x')dx' + \delta(x - x_0) \tag{8.5}$$

at a given point x^*. To this end, it is sufficient to replace φ^* in the form

$$\varphi^*(x^*) = (\varphi^*, \delta_{x^*}) = \int_X \varphi^*(x)\delta(x - x^*)dx. \tag{8.6}$$

It is assumed that x^* is a point of discontinuity of the function φ^*. We recall that for the pair of equations

$$\varphi = K\varphi + h, \quad \varphi^* = K^*\varphi^* + f$$

the following relation holds:

$$(\varphi^*, h) = (f, \varphi).$$

It follows from this and (8.6) that $\varphi^*(x^*) = \varphi(x_0)$ for $h(x) = \delta(x - x^*)$. Consequently, according to (8.4) we have

$$\varphi^*(x^*) = \mathbf{M}\xi^*_{x_0},$$

where

$$\xi^*_{x_0} = \delta(x - x^*) + \sum_{n=0}^{N} Q_n k(x_n, x^*). \tag{8.7}$$

The random variable (8.7) for the quantity $\varphi^*(x^*)$ is called a local estimate [8.2]. If in (8.5) instead of $\delta(x - x_0)$ an arbitrary function $f \in L_1$ is considered, then the local estimate can be constructed by the use of the randomization of (8.7) with respect to x_0 (Sect. 1.6).

Note that to construct the estimate of type (8.4) with zero variance in the case with alternating signs, it is necessary to use the importance function $\varphi - h$ [it may appear that this function varies weaker than φ, as is seen from the above example with $h(x) = \delta(x - x^*)$].

The local estimate for the intensity of radiation $\Phi(z, \mu)$ in a plane-parallel homogeneous layer is defined by (Sect. 2.4)

$$\sigma^{-1}k(z, \mu; z^*, \mu^*) = qw(\mu, \mu^*)\exp\{-\sigma(z^* - z)/\mu^*\}/\mu^*, \tag{8.8}$$

if $(z^* - z)\mu^* > 0$, otherwise $\sigma^{-1}k(z, \mu; z^*, \mu^*) = 0$. This is a function of the collision point (z, μ), i.e., of the coordinate of a particle immediately before the choice of the type of collision. We now assume that it is necessary to calculate the integral

$$\Phi_0(z^*) = \int_{-1}^{1} \Phi(z^*, \mu)d\mu.$$

We use the equality

$$h(z,\mu) = \delta(z - z^*)/\sigma.$$

Then

$$(Kh)(z,\mu) = q\int_A^B w(\mu,\mu')\frac{\exp\{-\sigma(z^* - z)/\mu'\}}{\mu'}d\mu', \tag{8.9}$$

where $(A, B) = (0, 1)$ for $z^* > z$, and $(A, B) = (-1, 0)$ for $z^* < z$.

The integral (8.9) can be calculated by the randomization principle using one random value of μ'. If these values are sampled independent of the history of the chain of the collision, then the variance increases compared to the case where (8.9) is calculated exactly. The expression of the randomized algorithm can be derived by the formula of total variance (Sect. 1.5). However to randomize (8.9), it is possible to use the value μ' obtained after the simulation of the scattering. In this case the scattering simulation scheme (Sect. 1.3.) is used. The corresponding estimate has the form

$$\xi_x = \sum_{n=0}^{N} Q_n h_0(z_n, \mu_n), \tag{8.10}$$

where $x = (z_0, \mu_0)$ is the point where the particles begin, and

$$h_0(z,\mu) = \exp\{-\sigma(z^* - z)/\mu\}/\mu \tag{8.11}$$

if $(z^* - z)\mu^* > 0$, otherwise $\sigma^{-1}k(z,\mu; z^*,\mu^*) = 0$.

It is also possible to use equations which are obtained by multiple application of the operator K to (8.1). For example, the double application gives the following estimate for $\varphi(x)$:

$$\xi_x^{(2)} = h(x) + (Kh)(x) + \sum_{n=0}^{N} Q_n(K^2h)(x_n). \tag{8.12}$$

The known double local estimate for $\varphi^*(x^*)$ can be obtained from this if we take $h(x) = \delta(x - x^*)$. Randomization of such an estimate in the transfer problems is usually carried out with respect to the directions and the free path length of an auxiliary path from the point r_n in a random auxiliary direction [8.2].

Further generalizations of (8.4) lead to the estimate

$$\xi_x^{(m)} = \sum_{n=0}^{m-1} K^n h|_x \sum_{n=0}^{N} Q_n(K^m h)(x_n). \tag{8.13}$$

In Sect. 8.3 we will study the variance of such estimates.

8.2 Translations

We now assume that it is necessary to calculate the solution to (8.1) at a set of points $x^{(1)}, ..., x^{(n)}$. The coordinate translations could be used to shift all these points to a fixed point x. As a result, we obtain a set of n equations which must be solved at the point x. The solutions can be obtained by the weighted Monte Carlo estimates constructed on a Markov chain.

It is clear that this method will be effective if the kernel of (8.1) has some kind of symmetry relative to the translation used. Therefore, we now describe this approach for a special optics problem of a calculation of the irradiance contrast. To be more specific, we use the terminology of the atmospheric optics problems. Differences in the optical properties of media are not important for the algorithm under study.

Let us consider a layer $0 \geq z \geq H$ of a 3-D space filled with a scattering and absorbing material. A flux $I_0(P,\omega)$ of photons is incident on the boundary $z = H$; ω is the direction of the flux, P is a point with coordinates (x_p, y_p, H) lying on the plane $z = H$. Interaction of the light with the material is defined by the extinction cross section $\sigma(r)$ after a collision, and the scattering indicatrix $w_s(\mu_s, r)$ which satisfies the condition

$$\int_{-1}^{1} w_s(\mu_s, r) d\mu = 1.$$

Here μ_s is the cosine of the scattering angle, the point $r = (x, y, z)$ is lies inside the layer. When the light having the direction ω is incident on the plane $z = 0$, it is reflected in a direction ω'. We denote by $A(Q, \omega, \omega')$ the irradiance coefficient which is defined as the ratio of the intensity of the light propagating from the point Q in the direction ω' to the irradiance of a unit area element situated at the point Q perpendicular to the direction ω. Here $Q = (x_Q, y_Q, 0)$ is a point lying in the plane $z = 0$.

In order to solve the problem of calculating the transfer function of the irradiance contrast, it is necessary to define the functions $I_0(P,\omega)$ and $A(Q,\omega,\omega')$ (i.e., to describe the initial contrast). Then the distribution (in the plane $z = H$) of the intensity of the light emanating from the layer or functions of this intensity are calculated by dependent samplings.

In calculations of photographic systems and photometric characteristics of luminescent screens the irradiance contrast is usually described by the source function $I_0(P,\omega)$. In this case one takes, as a rule, $H \to \infty$, or $A(Q,\omega,\omega') == 0$. To investigate the influence of the atmosphere on the irradiance contrast when observing the Earth's surface from space, it is necessary to describe the initial contrast by the function $A(Q,\omega,\omega')$. We will consider the problem of calculating the intensity of light emanating from the layer through the boundary $z = H$ in the direction ω^* at points $P_1^*, ..., P_n^*$, i.e., the vectors

$$I(P^*, \omega^*) = \{I(P_1^*, \omega^*), ..., I(P_n^*, \omega^*)\}$$

for

$$I_0(P,\omega) = I_0(P)\delta(\omega - \omega_0).$$

This problem has two specific features which must be incorporated in the constuction of the Monte Carlo algorithm. First, some values of the light intensity are calculated at fixed points for a given view direction.

Second, the dependence of the function $I(P^*,\omega^*)$ on the irradiance distribution of the initial contrast is calculated. The most difficult and practical problem is the evaluation of the dependence of $I(P^*,\omega^*)$ on small variations of the initial irradiance which lead to small variations of the function I.

(a) Use of dependent trajectories. The direct Monte Carlo scheme is not applicable here. Additional difficulties are caused by the infinite flux of photons. In this case it is useful to apply method of adjoint trajectories [8.5], in which the trajectories start at a point in the direction ω^* which is backward relative to the view direction ω^*. At every collision point the densities of the source of single scattered photons are summed.

To explain how the trajectories are constructed and clarify the calculation of the irradiance we consider an arbitrary trajectory

$$\{(r_0, -\omega^*, Q_0), (r_1, \omega_1, Q_1), ..., (r_k, \omega_k, Q_k)\}.$$

Here $r_i(x_i, y_i, z_i)$ is the position of the i-th collision point ($r_0 = P^*$), ω_i is the unit direction vector after the i-th collision ($Q_0 = 1$), r_k is the position of the k-th collision preceding the escape.

It is well-known that it is possible to improve an algorithm if the trajectories are simulated without escapes; the absorption of the photon is taken into account by the multiplication of the auxiliary weight by the survival probability $q(r)$ after a collision at point r. After the reflection at the plane $z = 0$ the weight is multiplied by

$$a(Q,\omega') = \int_{\Omega} A(Q,\omega,\omega') cos(\omega, n) d\omega,$$

where n is a unit vector normal to the plane $z = 0$ at the point Q which is directed inside the layer. The irradiance $I(P^*,\omega^*)$ is calculated by averaging (over all trajectories) the following random estimate:

$$\xi = \sum_{i=1}^{k} \frac{e^{-\tau_i}\varphi_i Q_i}{2\pi} I_0(P_i). \tag{8.14}$$

Here τ_i is the optical length measured along the direction $-\omega_0$ from the i-th collision to the point P_i lying on the plane $z = H$. The quantity φ_i is defined by

$$\varphi_i = w_s[(\omega_{i-1}, -\omega_0), r],$$

if the i-th collision occurred inside the layer at point r, and

$$\varphi_i = A(Q, -\omega_0, \omega_{i-1}) \cos(-\omega_0, n) 2\pi / a(Q, \omega_{i-1}),$$

if the photon reaches Q at the plane $z = 0$ (first, the weight, then φ_i are calculated).

(b) Use of the correlated estimates. The problem is to evaluate the dependence of the desired irradiance on the disposition of the observation point.

We suppose that it is necessary to calculate the irradiances $I_1 = I(P_1^*, \omega^*)$, $I_2 = I(P_2^*, \omega^*)$ at two arbitrary points $P_1^* = (x_1^*, y_1^*, H)$ and $P_2^* = (x_2^*, y_2^*, H)$ lying in the plane $z = H$. If we evaluate the values I_1, I_2 in two independent calculations with errors $\delta\hat{I}_1, \delta\hat{I}_2$ respectively, then it may appear that

$$\delta(\hat{I}_1 - \hat{I}_2) = (\delta^2 \hat{I}_1 + \delta^2 \hat{I}_2)^{1/2} > |I_1 - I_2|.$$

To overcome this difficulty, the method of correlated sampling can be used [8.2]. In this method, different variants of the estimates are constructed using the same set of random numbers $\{\alpha_n\}$. Then the correlation between $\hat{I}_1$ and $\hat{I}_2$ is positive. Thus, it may appear that the variance

$$\mathbf{D}(\hat{I}_1 - \hat{I}_2) = (\delta\hat{I}_1)^2 + (\delta\hat{I}_2)^2 - 2\rho(\hat{I}_1, \hat{I}_2)$$

would be less than the quantity $(\delta\hat{I}_1)^2 + (\delta\hat{I}_2)^2$ which corresponds to independent calculations. Here $\rho(\hat{I}_1, \hat{I}_2) \geq 0$ is the co-variance of $\hat{I}_1$ and $\hat{I}_2$.

The standard use of pseudorandom numbers "in succession" gives no possibility for constructing correlated samples, because of the possible differences in the length of the trajectories in different calculations. However, it is possible to divide beforehand the sequence $\{\alpha_n\}$ into large sub-sequences and use them to simulate the corresponing trajectories. This procedure is simple when the multiplicative congruental method [8.2]

$$u_0 = 1, \quad u_n = u_{n-1} 5^{2p+1} (\mathrm{mod} 2^m), \quad \alpha_n = u_n 2^{-m}$$

is used. In this method

$$\alpha'_n = \alpha_{kn} = \{(5^{(2p+1)k})^n 2^{-m}\},$$

i.e., the initial number of the parts of length k are also constructed by the multiplicative generator with the factor $5^{(2p+1)k}$, which is defined by the value of α_k (since only m lowest digits are used). Thus, the simulation of the trajectories in different variants begins with the same pseudonumbers which lead to the coincidence (or similarity) of significant parts of the trajectories.

The method just described is all purpose. It makes it possible to correlate the estimates of the irridiance $I(P_i^*, \omega^*)$ whose variation is due to the surface relief.

In the problem studied the reflected surface is the plane $z = 0$. Therefore, there exists a more effective approach in order to achieve the desired correlation which is based on the use of one and the same set of trajectories corresponding, e.g., to a point P_1^*. In this approach the problem of calculating the irradiance I_2 at a point P_2^* is replaced with an equivalent problem of calculating the irradiance $\tilde{I} = \tilde{I}(P_1^*, \omega^*)$ in a system with the characteristics $\tilde{I}_0, \tilde{A}, \tilde{\sigma}, \tilde{q}, \tilde{w}_s$, which are obtained from the original system by the translation

$$P_1^* - P_2^* = (x_1^* - x_2^*, y_1^* - y_2^*, 0).$$

We denote the point $P - P_1^* + P_2^*$ by $\tilde{P}$. Then, for example,

$$\tilde{A}(Q, \omega, \omega') = A(\tilde{Q}, \omega, \omega').$$

In accordance with the weighted methods, in order to obtain an unbiased estimate for the irradiance $I_2 = \tilde{I}_1$, it is necessary to recalculate the weight of the photon as follows. After each collision inside the layer, e.g., at a point $r(x, y, z)$, the weight must be multiplied by $q(r)$, and after a reflection at $Q = (x_Q, y_Q, 0)$ lying on the plane $z = 0$, the weight is multiplied by $A(\tilde{Q}, \omega, \omega')$. Then the free length $l = |r_i - r_{i+1}|$ is simulated, and the weight is multiplied by

$$\frac{\sigma(\tilde{r}_{i+1})}{\sigma(r_{i+1})} \exp\left\{ - \int_0^l [\sigma(\tilde{r}_i + t\omega_i) - \sigma(r_i + t\omega_i)] dt \right\}.$$

If r_{i+1} is the position of the photon lying on the plane $z = 0$, then the corresponding weight factor is equal to

$$\exp\left\{ - \int_0^{|r_{i+1} - r_i|} [\sigma(\tilde{r}_i + t\omega_i) - \sigma(r_i + t\omega_i)] dt \right\}.$$

Next the scattering angle is simulated at a collision point r_i, and the weight is multiplied by $w_s[\tilde{r}_i, (\omega_{i-1}, \omega_i)]/w_s[r_i, (\omega_{i-1}, \omega_i)]$.

The random quantity which determines the contribution of a trajectory to the irradiance has a form

$$\begin{aligned} \xi_1 &= \sum_{i=1}^{k} \frac{e^{-\tilde{\tau}_i} \tilde{\varphi}_i \tilde{Q}_i}{2\pi} I_0(\tilde{P}_k), \\ \tilde{\tau}_i &= \int_0^{(H - z_i)/(\omega_0, n)} \sigma(\tilde{r}_i - t\omega_0) dt. \end{aligned} \tag{8.15}$$

Here n is a unit vector normal to the boundary $z = H$. Then

$$\tilde{\varphi}_i = W_s[(\omega_{i-1}, -\omega_0), \tilde{r}_i]$$

when the collision occurs in the layer at the point r, and

$$\tilde{\varphi}_i = A(\tilde{Q}, -\omega_0, \omega_{i-1}) \cos(-\omega_0, n) 2\pi / a(\tilde{Q}, \omega_{i-1})$$

after the reflection of the photon at the point Q lying in the plane $z = 0$.

This algorithm is easily constructed in the case of a horizontally homogeneous medium. In the first variant of the dependent sampling method the trajectories with equal indices are all equal (for all observation points) to within a shift of coordinates, since the absorption on the plane $z = 0$ is taken into account by a change of the weight. Therefore, the two dependent sampling methods are numerically equivalent but the second is more effective, since all the estimates are calculated on one and the same set of trajectories.

The algorithm described was used to evaluate the boundary curve, the function which describes the diffusion of the boundary between two half-planes with different reflection coefficients in the plane $z = 0$. The following problem was solved. A plane flux of photons $I_0(P,\omega) = \delta(\omega - \omega_0)$ is incident on the upper bound of a homogeneous plane-parallel layer of the atmosphere with optical depth τ; $\nu = \arccos(-\omega_0, n)$ is the direction of the flux measured from the vector n, the exterior normal to the boundary. We choose the coordinates so that the axis z coincides with n, and the plane $z = 0$ is the ground surface whose albedo is

$$a(x) = \begin{cases} A_-, & x < 0, \\ A_+, & x \geq 0. \end{cases}$$

The photons reflect from the surface according to the Lambert reflection law which means that the density of the cosine of the reflection angle has the form $P(\mu) = 2\mu$.

The boundary curve $I_\tau(x)$ describes the behavior of the irradiance for fixed $\omega = \omega^*$ along the line in the plane $z = H$ which is perpendicular to the boundary between two reflecting half-planes. We are interested in the behavior of this function in the neighbourhood of the plane $z = 0$.

To evaluate the desired function we define in the plane $z = H$ the points $P_1^*, ..., P_n^*$ with coordinates $\{h(j - (n+1)/2), 0, H\}$ and calculate the values

$$I_\tau(x_j) = I[P_j^*, \omega^*].$$

Here $j = 1, .., n$ (n odd), h is the distance between two points.

Assume that the set of the trajectories is constructed for the point $P_{(n+1)/2}^* = (0, 0, H)$. As mentioned above, each trajectory gives a condition to the irradiance simultaneously for all observation points.

Denote the contribution from a certain trajectory to the irradiance at a point P_j^* by $\xi_j (j = 1, ..., n)$. In view of (8.14) the quantity ξ_j is calculated by

$$\xi_j = \sum_{i=1}^{k} \frac{1}{2\pi} \exp\left\{-\frac{H - z_i}{\cos\nu}\right\} \varphi_i Q_{ij}.$$

Here $\varphi_i = w_s(\omega_{i-1}, -\omega_0)$, $Q_{ij} = Q_{i-1,j}q$, if the i-th collision occurred in the atmosphere, and $\varphi_i = 2\cos\nu$, $Q_{ij} = Q_{ij}a[x_i + h(j - (n+1)/2)]$, if the photon was reflecting from the plane $z = 0$ at the point x_i.

Thus for all the observation points the corresponding contributions from the trajectories which did not reach the plane $z = 0$ are equal. The contributions from the trajectories of the photons reflecting from the plane $z = 0$ are also equal for the observation points for which the weights are multiplied after each reflection by the same reflection coefficient. In other cases the contributions for different observation points are, in general, different since the weight factors are not equal.

The results of calculation of the function $I_\tau(x)$ were obtained for the scattering indicatrix $w_s(\mu)$ for the wave length $\lambda = 0.55\mu$m [8.3]. The calculations were carried out for different values of the optical depth τ from 0.1 to 0.8 with the step 0.1 and for three different incident directions: $Q = 0°, 30°$, and $60°$. We also studied the dependence of $I_\tau(x)$ on the reflection angle A_-, A_+ and on the survival probability after a collision. The calculations were carried out for different combinations of values of A_-, A_+ varying from 0 to 0.8 and for six values of the survival probability $q = 1, 0.9, 0.8, 0.7, 0.4, 0.2$.

All the results were obtained for the vertical observation direction, i.e., for $n = 21, h = 0.1H$ (i.e., in the segment $[-H, H]$). For each value of τ about 300 variants of $I_\tau(x)$ were calculated using a set of 1000 trajectories. The computer time varied from 13 to 25 min depending on the values of τ. The corresponding error in the irradiance calculations varied from 1% to 4% [8.4].

8.3 Some Relations Between the Variances

Let us consider the variance of (8.13). If $\rho < 1, h \in L_\infty$, then

$$\|K^m\| \to 0,\ m > \infty. \tag{8.16}$$

Assuming that $\mathbf{D}\xi_x < \infty$ for $h \equiv \text{const}$, it follows that

$$\mathbf{D}\xi_x^{(m)} \to \infty, m \to \infty.$$

Note that the random term in (8.13) estimates the function

$$\varphi^{(m)}(x) = \varphi(x) - \sum_{n=0}^{m-1} K^n h|_x.$$

It may happen that this expression is close to a constant even if m is not large. Therefore, the direct simulation may provide a small variance of the estimate as in (8.13) for small m if $k(x, x')$ is substochastic.

Under well-known conditions (e.g., the operator K is positive and compact) the function

$$\varphi^{(m)} = \sum_{n=m}^{\infty} K^n h$$

is relatively close to the first eigenfunction φ^* of the operator K. This function, in the case of the transfer integral operator, is varies slowly for an optically thick homogeneous media; moreover, the convergence $\varphi^{(m)} \to c_m\varphi(x)$ is fast

(Sect. 5.4). However, the assumption that the averaging transformation always leads to the decrease of the variance is mistaken. It is not difficult to construct examples with

$$\mathbf{D}\xi_x^{(1)} > \mathbf{D}\xi_x.$$

We assume, e.g., that it is necessary to calculate the probability that a particle escapes the half-space $z > 0$ filled with a scattering material. It is supposed that the source of particles lies inside the half space, and the survival probability q is equal to one. In this case, the physical estimate obtained as a direct evaluation of the number of escaping particles has zero variance. The estimate $\xi_x^{(1)}$ is constructed on the basis of (8.9) where, under the integral sign, the additional factor $|\mu'|$ is introduced. This factor appears because that the probability that the particles escape through the boundary $z = 0$ is defined by the function h as described in Chap. 2. It is clear that $\mathbf{D}\xi_x^{(1)} > 0$.

If $q < 1$ in the upper example, then the variance of the physical estimate is not equal to zero. Moreover, it may exeed the quantity $\mathbf{D}\xi^{(1)}$. From this point of view the calculation of the effective fission coefficient k_{eff} [8.5] for the simplest homogeneous probabilistic model of a reactor is of interest. Two main schemes for the averaging method for calculating the probability of fission of the neutrons from a specific source are as follows.

(1) Rather than calculating the number of fissions, one caculates (after each collision) the probability that the neutron will generate a fission after the subsequent collision.

(2) As an estimate for the number of fissions, the sum of probabilities of fission after each collision is used (i.e., a probability of the type σ_f/σ_t, where σ_f is the fission cross section, σ_t is the total cross section).

To estimate approximately the effectiveness of (1) and (2), we consider a model critical system with $k_{\text{eff}} = 1$, and with constant f, c, s (f is the fission probability, c is the absorption probability, and s is the scattering probability) and with constant l, the probability of escaping ($f + c + s = 1$).

Let ϑ be the number of neutrons escaping after a fission. Since $k_{\text{eff}} = 1$, we get

$$\frac{f(1-l)}{l + c(1-l)} = \frac{1}{\nu - 1}. \tag{8.17}$$

We now present the variances of k_{eff} for (1) and (2).

Scheme (1). The random estimate has the form

$$\xi_1 = Nf(1-l),$$

where N is the random number of neutron runs in the system.

It is easy to show that the probabilistic distribution of N is given by

$$P(n) = (1-l)^{n-1}s^{n-1}[1-(1-l)s], \quad n = 1, 2, \dots \ .$$

Using the generating functions, we obtain

$$\mathbf{D}(N) = \frac{s(1-l)}{[1-(1-l)s]^2},$$

$$\mathbf{D}(\xi_1) = \frac{f^2 s(1-l)^3}{[1-(1-l)s]^2}.$$

The variance corresponding to the direct calculation of the number of fissions is

$$\mathbf{D}\xi = \frac{f(1-l)[c(1-l)+l]}{[1-(1-l)s]^2}.$$

It follows using (8.17), that

$$\frac{\mathbf{D}\xi}{\mathbf{D}\xi_1} = \frac{c(1-l)+1}{f(1-l)s(1-l)} > \frac{\vartheta - 1}{s(1-l)} > \vartheta - 1.$$

It is known that the cost of the Monte Carlo methods is defined as the product of the variance and the computing time for one trajectory. Scheme (1) requires a computing time which is 20 – 30% higher than that of the direct simulation scheme.

Scheme (2). As in scheme (1), we can show that

$$\mathbf{D}\xi_2 = \frac{f^2(1-l)[l+s(1-l)]}{[1-(1-l)s]^2}.$$

Consequently,

$$\frac{\mathbf{D}\xi_2}{\mathbf{D}\xi_1} = \frac{l+s(1-l)}{s+(1-l)^2} = \frac{1+ls^{-1}(1-l)^{-1}}{(1-l)} > 1.$$

The calculations show that the inequality $\mathbf{D}\xi_2/\mathbf{D}\xi_1 > 1$ is satisfied. However it is sometimes useful to apply scheme (2), since it does not need as much computing time.

8.4 Notions on the Functional Convergence of the Estimates

It is clear that the averaging transformations described in Sect. 8.1 may lead to an improvement of the functional convergence of the Monte Carlo estimates. Let us consider the local estimate (8.8) for the radiation intensity in a plane-parallel layer. The corresponding estimate ξ_x is differentiable with respect to μ^*, and the derivative is uniformly bounded in the interval $|\mu^*| \geq \varepsilon > 0$ by a random variable with a finite variance (provided that the simulated Markov chain provides the finiteness of the variance for $h \equiv \text{const}$). Consequently, by using (8.8), it is possible to obtain a functional estimation of the intensity $\Phi(z^*, \mu)$ at a fixed point z^* which satisfies Theorem 1.1 for $|\mu| > \varepsilon > 0$.

An analogous analysis of (8.10) for the integral flux of the particle

$$\Phi_0(z^*) = \int_{-1}^{1} \Phi(z^*, \mu) d\mu \quad ,$$

considered as a function of z^*, gives a negative result since the derivative of (8.10) with respect to z^* has a discontinuity at $z^* = z$ (and the function (8.10) as well).

The estimate (8.9) possesses the best functional properties; however, it also has a discontituity at $z^* = z$, since the following relations hold:

$$\lim_{z^* \downarrow z} \int_0^1 \left[w(\mu, \mu') \exp\left\{ -\frac{\sigma(z^* - z)/\mu'}{\mu'} \right\} \right] d\mu' = \infty,$$

$$\lim_{z^* \downarrow z} \int_{-1}^{0} \left[w(\mu, \mu') \frac{\exp\{-\sigma(z^* - z)/\mu'\}}{|\mu'|} \right] d\mu' = \infty \quad .$$

Thus, the local estimates described in Sect. 8.1 do not ensure the weak convergence (with respect to the metric C) in calculating the integral flux of the particle. Therefore, it is important to use the vector algorithm described in Sect. 5.3, which is based on the stratified sampling of z.

Note that the method in Sect. 8.2 for calculating the function

$$I(P, \omega^*) = \varphi^*(P)$$

provides the mentioned weak convergence if the albedo $a(Q, \omega)$ is a smooth function of the point Q.

9. Monte Carlo Methods and Perturbation Theory

A triangular system of integral equations obtained by a multiple differentiation with respect to a parameter of the original integral equation is considered. Weak conditions are presented under which the unique solution to the system is given by the vector of derivatives. The technique presented is also generalized on the calculation of perturbations of different order. Possible applications of the method are considered for solving stochastic, optimization and some inverse problems. To solve the triangular systems of integral equations, the vector Monte Carlo estimates are used. The variance of these estimates is finite if the standard variance finiteness criterion for the scalar "collision estimate" of the original integral equation holds. This is also the basis for investigating the functional convergence of the Monte Carlo methods. Vector estimates are considered for the derivatives (of the first and second order) of the linear functionals when a part of the medium is replaced with a vacuum. Finally, applications for a boundary value problem for the second order elliptic partial differential equation are considered.

9.1 Vector Weighted Monte Carlo Methods

Let us consider a system of integral equations of a special triangular form

$$\varphi_i(x) = \sum_{j=1}^{i} \int k_{ij}(x,y)\varphi_j(y)dy + h_i(y), \tag{9.1}$$

or in the operator form

$$\Phi = \mathbf{K}\Phi + H,$$

where $k_{ii}(x,y) \equiv k(x,y)$, i.e., the diagonal operators are identical. It is assumed that $\mathbf{K} \in [L_\infty \to L_\infty]$ and the spectral radius of $\mathbf{K}$ is less than unity: $\rho(\mathbf{K}) < 1$. To construct a vector estimate for the solution, a terminating Markov chain is defined as follows: $x_0 = x$ and $x_1, \ldots, x_N$ are constructed according to the transition density $p(x,y)$ which is not equal to zero on the support of the matrix kernel $K(x,y)$, i.e., on the union of the supports of the kernels $k_{ij}(x,y)$.

Random matrix weights are determined as follows:

$$Q_0 = \{\delta_{ij}\}, \quad Q_n = Q_{n-1}\frac{K(x_{n-1}, x_n)}{p(x_{n-1}, x_n)}, \quad n = 1, 2, ...,$$

where δ_{ij} is the Kronecker symbol. If the additional condition $\rho(\mathbf{K}_1) < 1$ (where $\mathbf{K}_1$ is obtained from $\mathbf{K}$ by replacing k_{ij} with $|k_{ij}|$) is valid, then the following equalities hold:

$$\mathbf{M}\xi_x = \Phi(x),$$

$$\xi_x = \sum_{n=0}^{N} Q_n H(x_n).$$

As shown in [9.1], the covariance matrix

$$\Psi(x) = \mathbf{M}(\xi_x \xi_x^T)$$

satisfies the following matrix-integral equation:

$$\Psi(x) = [H\Phi^T + \Phi H^T - HH^T]_x + \int_X \frac{K(x,y)\Phi(y)K^T(x,y)}{p(x,y)} dy$$

or

$$\Psi(x) = \mu + \mathbf{K}_p \Psi$$

provided the Neumann series for the equation

$$\Psi = \mu_1 + \mathbf{K}_{p,1}\Psi$$

converges; here the matrix-integral operator $\mathbf{K}_{p,1}$ and the function μ_1 correspond to the equation in which the functions are replaced with the moduli.

Denote by $K_{p,ij}$ the integral operator with the kernel

$$k_{ij}^2(x,y)/p(x,y).$$

It was shown in Sect. 5.5 that for the triangular system of integral equations [i.e., when $K(x,y)$ is a triangular matrix], the following inequality holds:

$$\rho(\mathbf{K}_p) \leq \max_i \{\rho(K_{p,ii})\}$$

provided all the operators $K_{p,ij}$ are bounded. Consequently, we can formulate

Lemma 9.1. If the operators with the kernels

$$k_{ij}^2(x,y)/p(x,y)$$

are bounded, then the relation

$$\rho(K_p) \leq \rho(K_p)$$

holds for systems of form (9.1).

As a special generalization of (9.1) we consider a family of integral equations

$$\varphi(x,\lambda) = \int\limits_X k(x,y,\lambda)\varphi(y,\lambda)dy + h(x,\lambda)$$

depending on a parameter λ. In this case, the Monte Carlo estimates are scalar values; however, they depend on λ, i.e.,

$$\xi_x = \xi_x(\lambda) = \sum_{n-0}^{N} Q_n(\lambda)h(x,\lambda),$$

where $Q_n(\lambda)$ are scalar weights.

A problem of functional convergence of the Monte Carlo estimates now arises. Obviously, it is directly related to the convergence of the random function

$$\sqrt{s}\left\{s^{-1}\sum_{i=1}^{s}\xi_x^{(i)}(\lambda) - \varphi(x,\lambda)\right\}, \quad \lambda_1 \le \lambda \le \lambda_2 \quad , \tag{9.2}$$

where $\xi_x^{(i)}(\lambda)$ are independent realizations of the random variable $\xi_x(\lambda), i = 1, ..., s$.

According to [9.2], function (9.2) is weakly convergent to a continuous Gaussion random function in the interval $\lambda_1 < \lambda < \lambda_2$, if the following conditions hold:

$$|\frac{\partial \xi_x(\lambda)}{\partial \lambda}| \le G_x, \quad \mathbf{M}G_x^2 < +\infty, \quad \lambda_1 < \lambda < \lambda_2. \tag{9.3}$$

The finiteness of $\mathbf{M}G_x^2$ can sometimes be derived from a special triangular system of integral equations (Sect. 9.2).

9.2 Differentiation of Integral Equations with Respect to a Parameter

Monte Carlo methods can be applied in solving problems of applied mathematics, which can be posed in the form of integral equations of the second kind or systems of such equations. In these equations, the integration can be carried out over a manifold whose dimension is less than that of the problem considered. This means that the integration is performed, in fact, over a surface lying in the phase space; thus the kernels include special generalized factors similar to the "δ-function".

There is no effective use for these equations in constructing conventional discrete algebraic methods. However, they permit the construction of Monte Carlo estimates on the basis of appropriate Markov chains.

Integral equations solved by the Monte Carlo methods can be divided in two classes. The first includes equations whose kernels are transition densities of the considered Markov chains. The second includes equations constructed by transformations of the original mathematical model which may not have a probabilistic interpretation. Equations of the transfer theory (first class) are considered in Sect. 9.3. In Sect. 9.4, we present a typical example of the equation which belongs to the second class and relates to the Dirichlet problem for the Helmholtz equation.

The kernels of the integral equations often depend on parameters describing certain variable characteristics of the original information, for example, properties of the medium where the process is considered. In this connection, various problems, for example, inverse problems of prediction of the real values of parameters, can be considered using the observation of some functionals of the solution to (9.1). In addition, optimization problems with respect to parameters for these functionals are considered. When the parameters are random, we also consider the problem of calculating perturbations of the solution or the functionals under perturbations of the parameters. Sometimes, it is more effective to calculate derivatives of the solution (or functionals) with respect to parameters. We present a special vector technique for calculating the derivatives and the perturbations mentioned.

Consider in $L_\infty(X)$ an integral equation of the second kind with a kernel depending on a parameter λ:

$$\varphi(x,\lambda) = \int_X k(x,y,\lambda)\varphi(y,\lambda)dy + h(x,\lambda). \tag{9.4}$$

We rewrite (9.4) in the operator form $\varphi = K\varphi + h$. In (9.4), X is an n-dimensional Euclidean space. Consider the problem of calculating the derivatives

$$\varphi^{(n)}(x,\lambda) = \frac{\partial^n \varphi(x,\lambda)}{\partial \lambda^n}.$$

We introduce the following notation: $K^{(n)}$ is the integral operator with the kernel $k^{(n)}(x,y,\lambda)$, and $\rho(K)$ is the spectral radius of the operator K. It is assumed that the functions $k^{(n)}$ and $h^{(n)}$ are measurable with respect to x. By formally differentiating (9.4) n-times with respect to λ, we obtain a triangular system of integral equations

$$\varphi^{(n)} = \sum_{i=0}^{n} C_n^i K^{(n-1)}\varphi^{(i)} + h^{(n)}, \quad n = 0, 1, ..., m, \tag{9.5}$$

or in the operator form $\Phi = K\Phi + H$.

Now we give a justification of the use of (9.5) for calculating the derivatives $\varphi^{(n)}$.

We introduce the additional notation that $F_m(f)$ is the column vector consisting of the derivatives of the function $f(\lambda)$ with respect to λ, i.e.,

$$F_m(f) = (f, f^{(n)}, ..., f^{(m)})^T;$$

$D_m(f) = \{d_{ni}\}$ is the lower triangular $(m+1)$-dimensional matrix with entries

$$d_{ni} = C_n^{(n-i)} f^{(n-i)}, i = 0, 1, ..., n; n = 0, ..., m.$$

Using the Leibnitz formula for the m-th derivative of the product of two functions, we obtain the following lemma.

Lemma 9.2. The following relation holds:

$$F_m\left(\prod_{i=0}^{k} f_i\right) = D_m(f_k)...D_m(f_1)F_m(f_0). \tag{9.6}$$

Theorem 9.1. Suppose that the inequalities

$$|k^{(n)}(x, y; \lambda)| \leq k_n(x, y), \quad |h^{(n)}(x, \lambda)| \leq h_0(x), \quad h_0 \in L_\infty$$

hold in the interval $\lambda - \varepsilon < \lambda < \lambda + \varepsilon$ for some $\varepsilon > 0$, the integral operators K_n with the kernels $k_n(x, y)$ are bounded, $n = 0, ..., m$ and $\rho(K_0) < 1$. Then $\rho(K) < 1$ and the functions $\varphi^{(n)}$ satisfy (9.5), $n = 0, 1, ..., m$.

Proof. It follows from the theorem in Sect. 5.5 that the relation $\rho(\mathbf{K}) \leq \rho(K_0) = \rho_0 < 1$ holds in the interval $(\lambda - \varepsilon, \lambda + \varepsilon)$. By Lemma 9.2, the corresponding Neumann series for (9.5) coincides with the series obtained by the formal differentiation of the Neumann series for (9.4). It follows from (9.6) that this series is uniformly majorized in $(\lambda - \varepsilon, \lambda + \varepsilon)$. The Neumann series for the majorant is finite, thus proving Theorem 9.1. □

Lemma 9.2 and Theorem 9.1 can be directly extended to the case of a vector parameter $\lambda = (\lambda_1, ..., \lambda_k)$, since the corresponding systems of type (9.5) can also be written in a triangular form with the diagonal entries coinciding with K. To do this, it is sufficient to choose an index of the functions $\varphi^{(i_1,...,i_k)}$ according to the natural order in which they are written in the equations. This kind of index is universal provided that the equations are written in accordance with the natural increasing order of the derivatives (note that inside a fixed order of the derivative the index can be arbitrary).

Let us now mention some problems where this technique for calculating derivatives can be applied. Suppose that σ is a random vector and the functionals $J_k(\sigma)$ which we study depend on $\sigma_i (i = 1, ..., r)$ almost linearly. Note that the last assumption can be considered in the neighbourhood of the point $\sigma^{(0)} = M\sigma$ (the probability of hitting this neighbourhood is close to one). Then the correlation moments of the vector $J = \{J_k(\sigma)\}$ and the autocorrelation moments of the vectors σ and J can be calculated by the linearization of $J(\sigma)$ using the Monte Carlo evaluations of the derivatives $\partial J_k/\partial \sigma_i$. Note that this calculation technique is developed in [9.1] for the transfer theory problems for the case of the piece-wise constant random scattering cross section σ. The calculation of the higher order derivatives can thus be used here to control the numerical results and to make them more accurate. In [9.1] we also present also examples of iterative numerical solution of an inverse problem of finding σ by linearizing the equation

$$J_k(\sigma) = J_k^{(0)}, \quad k = 1, ..., n \geq r,$$

where $J_k^{(0)}$ are known (measured) values of functionals.

Consider now a possibility of using estimates of the derivatives for solving special nonlinear integral equations.

Suppose that the parameter σ depends on the solution, i.e., $\sigma = S(\varphi)$; σ thus satisfies the following nonlinear equation:

$$\sigma = S[\varphi(\sigma, \cdot)]. \tag{9.7}$$

Linearizing $\varphi(\sigma, \cdot)$ in (9.7) with respect to σ at the point $\sigma = \sigma^0$, we obtain a simpler equation

$$\sigma = S[\varphi(\sigma^{(0)}, \cdot)] + \varphi'_\sigma(\sigma^{(0)}, \cdot)(\sigma - \sigma^{(0)})]. \tag{9.8}$$

The solution of this equation gives the first approximation $\sigma = \sigma^{(1)}$. Relation (9.8) thus determines an iteration procedure for solving (9.7) in a partly exact manner. Note that this procedure can be used to solve the heat transfer equation. In addition, calculation of higher order derivatives can also be used in making numerical results more accurate.

We consider a problem of calculating the derivative with respect to some direction ω, i.e., of the quantity

$$\varphi'(x) = \left.\frac{d\varphi(x + t\omega)}{dt}\right|_{t=0} .$$

To this end, we differentiate with respect to t the relation

$$\varphi(x + t\omega) = \int_X k(x + t\omega, y)\varphi(y)dy + h(x + t\omega). \tag{9.9}$$

Interchanging the order of differentiation and integration, we obtain

$$\varphi'(x) = \int_X k'(x, y)\varphi(y)dy + h'(x), \tag{9.10}$$

where

$$k'(x, y) = \left.\frac{dk(x + t\omega, y)}{dt}\right|_{t=0} .$$

This interchanging is additionally justified in all particular cases. Using standard Monte Carlo algorithms for solving (9.10), it is not difficult to construct probabilistic estimates for the quantity $\varphi'(x)$ [9.3].

However, in the radiative transfer theory which is an important area for applications of the Monte Carlo methods, this technique fails because the kernel of the integral equation involves a "σ-factor" which connects the new collision point with the direction of the particles (Sect. 9.3).

Therefore, it is useful to make the change of variables $x + t\omega \to x$; then (9.9) is transformed to the following equation:

$$\Psi(x,t) = \int_X k_1(x,y,t)\Psi(y,t)dy + h_1(x,t), \tag{9.11}$$

where

$$\Psi(x,t) = \varphi(x+t\omega), \quad k_1(x,y,t) = k(x+t\omega, y+t\omega), \quad h_1(x,t) = h(x+t\omega).$$

This change of variable is convenient in calculations if the original equation is invariant (at least partly) under the translation along the direction ω. Since the variable t is a parameter of (9.11), the construction of the triangular system of integral equations for the derivatives

$$\left.\frac{d^n\Psi}{dt^n}\right|_{t=0} = \left.\frac{d^n\varphi}{dt^n}\right|_{t=0}$$

can be carried out in an obvious manner.

We now consider the Monte Carlo algorithms for solving the constructed triangular system of integral equations. Lemma 9.1 shows that the variances of these algorithms are bounded if the operators with the kernals ${k^{(n)}}^2(x,y,\lambda)/p(x,y)$ are bounded and the spectral radius of the operator with the kernel $k^2(x,y,\lambda)/p(x,y)$ is less than unity ($n = 1,...,m$), i.e., if the standard variance finiteness condition for the scalar Monte Carlo estimates for equation (9.4) is satisfied.

To study the functional convergence of the Monte Carlo estimates to $\varphi(x,y)$ in some interval $\lambda_1 < \lambda < \lambda_2$ (Sect. 9.1), it is sufficient to consider an auxiliary triangular system of two integral equations of type (9.5), where $k(x,y,\lambda)$ is replaced with $|k(x,y,\lambda)|$, while $k(x,y,\lambda)$ is replaced with $|k'(x,y,\lambda)|$. This permits the construction of the majorant mentioned in the condition for (9.3).

Theorem 9.2. Suppose that the inequalities $|k(x,y,\lambda)| \le k_0(x,y)$ and $|k'(x,y,\lambda)| \le k_1(x,y)$ hold in the interval $\lambda_1 < \lambda < \lambda_2$. If K_1, the integral operator with the kernel $k_1^2(x,y)/p(x,y)$, is bounded and the spectral radius of $K_{p,0}$, the operator with the kernel $k_0^2(x,y)/p(x,y)$, is less than unity, then (9.3) is satisfied.

Remark. Note that the condition $\rho(K_{p,0}) < 1$ can be replaced by a wider assumption, i.e., the variance of the estimate $\xi_{x,0}$, corresponding to the operator K_0, is finite.

9.3 Calculation of Perturbations

We will now obtain a system of equations describing small perturbations of the solution of integral equations of the second kind under small perturbations of the kernel and the right-hand side. We will consider the two equations

$$\varphi = K\varphi + h, \quad \varphi_1 = K_1\varphi_1 + h_1.$$

Subtracting yields

$$\varphi_1 - \varphi = K_1(\varphi_1 - \varphi) + (K_1 - K)\varphi + h_1 - h.$$

If we neglect the term $(K_1 - K)(\varphi_1 - \varphi)$, then we obtain the following system of equations:

$$\varphi = K\varphi + h, \quad \delta\varphi = K(\delta\varphi) + \delta K\varphi + \delta h. \tag{9.12}$$

It is not difficult to see that (9.12) gives the principal part of the perturbation of φ if the dependence of the functional characteristics of the problem on the varying parameter is smooth enough. Let us suppose for simplicity that $\delta h \equiv 0$. We denote by $R(K)$ the resolvent of the operator K, i.e., we set $R(K) = (I - K)^{-1}$. From (9.12) we have

$$\delta\varphi = R(K)\delta K R(K)h.$$

This representation of small perturbations of the first order corresponds to the general perturbation theory [9.1] which gives the following obviously verified relation:

$$R(K_1)h = \sum_{n=0}^{\infty}[R(K)\delta K]^n R(K)h \tag{9.13}$$,

provided $\rho[R(K)\delta K] < 1$.

It is not difficult to see that the terms of (9.13) satisfy the following triangular system of integral equations:

$$\varphi_0 = K\varphi_0 + h, \quad \varphi_n = K\varphi_n + \delta K\varphi_{n-1}, \quad n = 1, 2, \ldots \ . \tag{9.14}$$

We shall now consider finite systems of type (9.13), i.e., $n = 1, ..., m$. In the next section we shall describe nonbiased vector weighted estimates for solving systems of integral equations. Note that this technique fails if the kernels of the operators K and K_1 are mutually singular, for example, if they include "δ-functions" which define integration over different manifolds whose dimentions are less than that of X. In this case, it is perhaps more appropriate to use the following system instead of (9.14):

$$\begin{aligned} &\varphi_0 = K\varphi_0 + h \\ &\varphi_n = \sum_{i=1}^{n}(\delta K)^{n-i}K\varphi_i + (\delta K)^n h; \quad n = 1, 2, ..., \end{aligned} \tag{9.15}$$

which is obtained by substituting the last expression for φ_n into (9.14), $n = 0, 1, \dots$.

Using (9.14, 15), one can sometimes calculate the solution of the integral equation $\varphi = K_1\varphi + h$ by the Monte Carlo method even if $\rho(K_1) = 1$, i.e., when the standard Monte Carlo estimates are not applicable.

To do this, it is necessary to construct an operator K, such that

$$\rho(K) < 1, \quad \rho(R(K)\delta K) < 1, \quad \rho(K^{(0)}) < 1, \tag{9.16}$$

where $K^{(0)}$ is the operator with the kernel $|k(x, y)|$. From (9.13), we obtain

$$\varphi = \lim_{n\to\infty} \Psi_n, \quad \Psi_n = \sum_{k=0}^{n} \varphi_k.$$

The successive summation of (9.14) gives the following system of integral equations for Ψ_n:

$$\Psi_0 = K\Psi_0 + h, \quad \Psi_n = K\Psi_n + \delta K\Psi_{n-1} + h, \quad n = 1, 2, \dots.$$

To solve finite systems of this kind, one can use vector Monte Carlo algorithms (Sect. 9.1) if conditions in (9.16) are satisfied.

The optimization of the method presented for the calculation of the function φ consists of obtaining a relation between the size of the system of equations and the size of a sample, which is such that the mean number of operations needed for obtaining the result to within a prescribed accuracy is minimal. Note that this optimization problem can be solved in a similar manner to the optimization of the Monte Carlo algorithms for the numerical realization of resolvent iterative processes [5.5].

We now consider the construction of the vector algorithms for (9.14). Representing the matrix weight factor as a matrix multiplied by the kernel $k(x, y)$, we obtain the following relations for evaluating the matrix weights:

$$Q_n = q_n Q'_n, q_0 = 1, q_n = q_{n-1} k(x_{n-1}, x_n)/p(x_{n-1}, x_n)$$

$$Q'_0 = Q_0 = \{\delta_{ij}\}, Q'_n = Q_{n-1}\Delta[1, \delta K(x_{n-1}, x_n)/k(x_{n-1}, x_n)] \quad ,$$

where $\Delta[1, a]$ is a bidiagonal matrix with the unities as the diagonal entries; the entries of the subdiagonal are equal to a. Consequently, only two different entries compose two nonzero diagonals of the matrix Q'_n thus essentially decreasing the cost of the algorithm.

It is not difficult to see that the following representation holds if $\delta h \equiv 0$

$$\varphi_1(x) = M\eta_x, \quad \eta_x = \sum_{n=1}^{N} q_n h(x_n) \sum_{m=0}^{n-1} \frac{\delta k(x_m, x_{m+1})}{k(x_m, x_{m+1})} \tag{9.17}.$$

The expression for η_x can be formally obtained from the relation

$$\delta q_n = q_n \delta \ln q_n.$$

It is interesting to note that this procedure can ensure a satisfactory estimation of the perturbations even if $\delta k/k$ is not small everywhere, provided the conditions in (9.16) hold.

9.4 Calculation of Derivatives

The kernel of the general one-group integral equation of the transfer theory for the collision "importance" $\varphi(x)$ in the phase space $X = (R \times \Omega)$ has the form [9.1]

$$k(x,x') = \frac{q(r)}{2\pi|r'-r|^2} g\left(\omega, \frac{r'-r}{|r'-r|}\right) \times \sigma(r') e^{-\tau(r,r')} \delta\left(\omega' - \frac{r'-r}{|r'-r|}\right).$$

Here, σ is the total scattering cross section, g is the scattering phase function, $\tau(r,r')$ is the optical length of the segment $(r,r'), q(r')$ is the survival probability of the radiation quantum undergoing a collision; the δ-function shows that the spatial point r' lies on the ray $r(t) = r + t\omega, t \geq 0$. Suppose now that $\sigma_s(r)$ takes constant values σ_1 and σ_2 on D_1 and D_2, respectively, where D_1 and D_2 are two disjoint spatial domains. Using the expression for optical length

$$\tau(r,r') = \int_0^{|r'-r|} \sigma(r(t))dt,$$

we obtain

$$\frac{\partial k(x,x')}{\partial \sigma_i} = k(x,x')b_i, \quad b_i = \left[\frac{\delta_i(r)}{\sigma_i} - \frac{\delta_i(r)}{\sigma(r)} + \frac{\delta_i(r')}{\sigma(r')} - \Delta_i l\right],$$

$$\frac{\partial^2 k(x,x')}{\partial \sigma_i \partial \sigma_j} = k(x,x')\left[b_i b_j - \frac{\delta_i(r)\delta_j(r)}{\sigma_i\sigma_j} + \frac{\delta_i(r)\delta_j(r)}{\sigma^2(r)} + \frac{\delta_i(r')\delta_j(r')}{\sigma^2(r')}\right] \tag{9.18}$$

where $i,j = 1,2$, the quantity $\delta_i(r)$ is the indication function of the domain D_i and $\Delta_i l$ is the length of the part of the ray $r(t)$ lying in D_i. The inequality $\rho(K) < 1$ in the case of the transfer equation considered follows from the inequality $\sigma_s(r)/\sigma(r) \leq q < 1$ or from the boundedness of the medium which leads to the fact that the particle escapes the medium with probability one [9.1]. If one of these conditions holds, then the integral operators with the kernels of type (9.18) are bounded and $\rho(K) < q_0 < 1$ if $\sigma_i < c < +\infty$. In addition, for all values of σ_1' and σ_2' lying inside the intervals $\sigma_1 \pm \varepsilon$, $\sigma_2 \pm \varepsilon$, respectively, with ε being sufficiently small, the following inequality holds:

$$k(x,x') \leq (1 + c_0\varepsilon)k_\varepsilon(x,x') \quad ,$$

where $k_\varepsilon(x,x')$ is the kernel of the transfer equation corresponding to $\sigma_i' - \sigma_i - -\varepsilon$, $i = 1,2$. By Theorem 9.1, it is thus possible to use the triangular systems of the integral equations considered above to construct the Monte Carlo estimates for the derivatives of the solution with respect to σ_1 and σ_2. Lemma 9.2 shows that the variances of such estimates are finite, for example, if the direct simulation of the transfer process is carried out, i.e., when $p(x, x') = k(x, x')$. It is then not difficult to see that the estimate of the first derivative coincides with that obtained by the differentiation of the estimate for the solution [1.8]. However, the vector technique permits one to obtain the most general condition under which the estimates are unbiased and have finite variances. In addition, the technique provides a possibility for constructing and investigating estimates for higher order derivatives.

We now consider the functional convergence of the estimates $\xi_x(\sigma_s)$ for the function $\varphi(x) = \varphi(x, \sigma_s)$, where σ_s is the scattering cross section in a domain D where $\sigma_s^{(1)} \le \sigma_s \le \sigma_s^{(2)}$. It is obvious that

$$\frac{k(x, y, \sigma_s)}{k(x, y, \sigma_s^{(1)})} \le \begin{cases} \sigma_s^{(2)}/\sigma_s^{(1)}, & y \in D \\ 1, & y \in D. \end{cases}$$

Consequently, we can set

$$k_0(x, y) = \frac{\sigma_s^{(2)}}{\sigma_s^{(1)}} k(x, y, \sigma_s^{(1)}).$$

By Theorem 9.2 (considering the corresponding remark), the functional convergence exists if $p(x, y)$ is chosen such that $D\xi_{x,0} < +\infty$. This is the case, for example, if the direct simulation without absorption for the operator K_0 is carried out and the following inequality holds:

$$\frac{\sigma_s^{(2)}}{\sigma_s^{(1)}} \sup_r q(r) < 1.$$

The conditions obtained are substantially extended if the domain D is not large. Moreover, the corresponding extension can be obtained using geometric arguments.

Note that the calculation of the derivatives with respect to $\sigma_c = \sigma - \sigma_s$ (σ_c is the absorption coefficient in D) is simpler as compared to the case of the scattering coefficient. For example, if the direct simulation is carried out without absorption in the domain D (i.e., $p(x, y)$ is obtained from $k(x, y)$ by replacing σ with σ_s in D), then the weight at the trajectory points not lying in D is calculated as $\exp(-L\sigma_c)$, where L is the trajectory length inside the domain D [9.1]. The variances of the estimates for the derivatives are then finite and the functional convergence holds if $\sigma_c \ge 0$.

Note that vector Monte Carlo algorithms [9.1] are applied in solving systems of transfer equations taking into account polarization as well as the multi-group system of equations of transfer. The technique presented is obviously generalized on these cases by using block triangular systems of integral equations which are obtained by differentiating original systems.

9.5 Calculation of Perturbations in the Transfer Theory

We now consider an example for evaluating perturbations in the transfer theory by the Monte Carlo method.

Suppose that a kernel $k_1(x,x')$ is obtained from the kernel $k(x,x')$ by replacing the cross section $\sigma(r)$ with zero in a spatial domain D. Then

$$\sigma k(x,x') = \begin{cases} -k(x,x'), & x' \in D \\ 0, & x' \in D. \end{cases}$$

Consequently, from (9.6) for $\sigma h \equiv 0$, we have

$$\varphi_1(x) = M\eta_x, \quad \eta_x = \sum_{n=1}^{N} q_n h(x_n) \sum_{m=0}^{n-1} \sigma_D(x_{m+1}) \quad , \tag{9.19}$$

where $\sigma_D(x)$ is the indication function of the domain D. For the direct simulation [i.e., when $p(x,y) \equiv k(x,y)$] we have

$$\eta_x = -\sum_{n=1}^{N} h(x_n) \sum_{m=0}^{n-1} \sigma_D(x_m).$$

A change in the order of summation in (9.19) yields

$$\eta_x = -\sum_{m=1}^{N} \sigma_D(x_m) \sum_{n=m}^{N} Q_n(x_n).$$

Using the iterated averaging and the collision estimate technique [1.7], we can show [11] that

$$M\eta_x = -\int_{\Omega}\int_{D} \varphi^*(x')\varphi(x')dx' \quad ,$$

where $\varphi^*(x')$ is the collision density.

It is interesting to note that the calculation technique according to (9.19) substantially differs from the Monte Carlo algorithm [9.4] based on the known formula of the small perturbation theory for the integro-differential transfer equation. In [9.4], it was noted that in order to realize the formula of the perturbation theory accurately, it is necessary to randomize the standard estimates by simulating additional particles.

An important property of the algorithm presented is the above-mentioned possibility of decreasing the error of the perturbation calculations on the basis of (9.14).

In the example considered, the matrix operations are particularly simple, because they are applied only to matrices of the type $\Delta(1,-1)$ and $\Delta(1,0) = \{\delta_{ij}\}$ (Sect. 9.3). Condition (9.16) can be verified on the basis of the inequality

$$\rho[R(K)\delta K] \le \|\delta K\|(1 - \|K\|)^{-1}.$$

In the transfer theory, the quantities $\|K\|$ and $\|\delta K\|$ are easily estimated using the medium constants [9.1].

Note that the transition density $p(x, x')$ must be chosen such that the probability of the occurrence of at least one collision in the perturbation region of D will not be too small. If the domain D has a small volume but a sufficiently large extension in space, then intersections of D without a collision often occur, i.e., collisions inside D rarely occur. It is then convenient to set

$$p(x, x') = k(x, x', t) \quad ,$$

where $k(x, x', t)$ is obtained from $k(x, x')$ by replacing in D the quantity σ with σt, where $t > 1$. The cost of the corresponding weighted algorithm is minimized by an appropriate choice of t.

9.6 Calculation of Derivatives of Solutions to Boundary Value Problems by the Monte Carlo Method

Consider the following three-dimentional Dirichlet problem:

$$\Delta u - cu = -g, \quad u \mid_{\Gamma} = \Psi \tag{9.20}$$

in a domain D, where Γ is the boundary of D, c is a non-negative constant. It is assumed that g, Ψ and Γ are regular enough so that there exists a unique solution to (9.20) and the following well-known Green formula for a sphere holds:

$$\begin{aligned} u(P_0) = & \frac{d_0\sqrt{c}}{4\pi d_0^2 \sinh(d_0\sqrt{c})} \int\limits_{S(P_0)} u(s)ds \\ & + \int_{|r-P_0|<d_0} \frac{\sinh[(d_0 - |r - P_0|)\sqrt{c}]}{4\pi|r - P_0| \sinh(d_0\sqrt{c})} g(r)dr, \end{aligned} \tag{9.21}$$

where $S(P_0)$, a sphere of radius $d(P_0)$ with the center at $P_0 \in D$, is the boundary of the sphere, $d_0 = d(P_0)$ is the distance from P_0 to the boundary Γ. The first integral in (9.21) is taken over the sphere $S(P_0)$ tangent to the boundary Γ.

Equation (9.21) can be considered to be an integral equation of the second kind with a generalized kernel. It makes it possible to consider the first integral in (9.21) as a three-dimensional one. It is not difficulat to see that standard Monte Carlo algorithms are applicable here if the kernel is used to construct the transition density of the Markov chain simulated. In our case the transition is carried out from the point P_0 to sphere $S(P_0)$, i.e., the "walk on spheres" is simulated.

Note that it is necessary to consider (9.21) together with the relation

$$u(P_0) = \Psi(P_0), \quad P_0 \in \Gamma$$

which means that the kernel of the equation vanishes if the first variable P belongs to Γ. Therefore, after the chain reaches the boundary, the walk terminates and the quantity $\Psi(P)$ (multiplied by a weight) is stored. However, this probabilistic scheme cannot be realized numerically, because the "walk on spheres" process with probability one does not reach the boundary after a finite number of steps. This is due to the L_∞-norm of the integral operator which is equal to unity. Below we consider an unbiased but numerically realizable Monte Carlo estimate [9.4].

We temporarily suppose that the solution to the Dirichlet problem is known at every point of the set

$$\Gamma_\varepsilon = \{P \in D \cup \Gamma : d(P) < \varepsilon\}.$$

Then $u(r)$ satisfies the following integral equation:

$$u(r) = \int\limits_D k(r,r')u(r')dr' + h(r) \quad , \tag{9.22}$$

where

$$k(r,r') = \begin{cases} \frac{d\sqrt{c}}{\sinh(d\sqrt{c})}\delta_r(r'), & r \in \Gamma_\varepsilon \\ 0, & r \in \Gamma_\varepsilon, \end{cases}$$

$$h(r) = \begin{cases} \frac{1}{4\pi} \int\limits_{|r-P_0|<d_0} \frac{\sinh[(d-|r'-r|)\sqrt{c}]}{|r'-r|\sinh(d\sqrt{c})} g(r')dr', & r \in \Gamma_\varepsilon \\ u(r), & r \in \Gamma_\varepsilon. \end{cases}$$

Here, $d = d(r)$, and $\delta_r(r')$ is the generalized density corresponding to the uniform probability distribution on the sphere $S(r)$. It is not difficult to show (Sect. 1.4) that in L_∞-space we have

$$\|K^2\| \le 1 - \nu(\varepsilon) < 1, \quad \nu(\varepsilon) = \varepsilon^2/(4d_{\max}^2).$$

This ensures the convergence of the Neumann series for (9.22); the Monte Carlo method is thus applicable. In addition, it implies that the solution to the original differential problem is a unique solution to (9.22). The generalized solution is also included. Therefore, to calculate the solution, the following relation can be used:

$$u(P_0) = M\xi, \quad \xi = h(P_0) + \sum_{n=1}^{N} Q_n h(P_n). \tag{9.23}$$

Here, $\{P_n\}$ is the "walk on spheres" process terminating in Γ_ε which can be called an ε-spherical process when $\varepsilon > 0$. The weights are defined by the formulae

$$Q_0 = 1, \quad Q_n = Q_{n-1}\frac{d_{n-1}\sqrt{c}}{\sinh(d_{n-1}\sqrt{c})} \le Q_{n-1},$$

$$d_n = d(P_n), \quad n = 1, 2, \ldots \quad .$$

Now we recall that the unreal assumption was made under which $u(r)$ is known in Γ_ε. However, instead of explicit values of $u(r)$ in Γ_ε, it is possible to use approximate values, for example, replacing them with values of Ψ at the nearest points of Γ, i.e., we can take

$$u(r) \approx \Psi(r^*), r \in \Gamma_\varepsilon, r^* \in \Gamma, |r - r^*| = d(r).$$

As a result, we obtain a biased estimate ξ_ε such that the difference between $u(P_0)$ and the expectation of ξ_ε is of order ε. Indeed, using $Q_N \leq 1$, we have

$$|u - u_\varepsilon| < |M\{Q_N[u(P_N) - \Psi(P_N^*)]\}| \leq A\varepsilon,$$

where A is a constant which is finite because the derivatives of $u(x)$ are finite in D.

The variance of the estimate constructed is finite because the direct simulation technique is realized for (9.22) with $c = 0$; while $Q_N < 1$ when $c > 0$. The variance of ξ_ε is bounded uniformly with respect to ε.

We now consider a method for estimating the derivatives of the solution to (9.20) with respect to the parameter c. To realize the technique described in Section 9.2, it is sufficient to calculate the derivatives of the factor $d\sqrt{c}/\sinh(d\sqrt{c})$ of the kernel of (9.22). In particular, the corresponding first derivative is determined by the expression

$$A(c) = \frac{d}{2\sqrt{c}} \frac{1}{\sinh^2(d\sqrt{c})}[\sinh(d\sqrt{c}) - d\sqrt{c}\cosh(d\sqrt{c})].$$

Using the relations

$$\sinh(x) = x + \frac{x^3}{6} + O(x^3), \quad \cosh(x) = 1 + \frac{x^2}{2} + O(x^2),$$

we obtain

$$\lim_{\varepsilon \to 0} A(c) = -\frac{1}{6}d^2.$$

On the other hand, it is obvious that

$$\lim_{\varepsilon \to \infty} A(c) = 0.$$

Consequently, the variance of the estimate for the derivative is bounded uniformly with respect to c if $\varepsilon > 0$. In addition, it follows from Theorem 9.2 that there is a functional convergence of the estimate for $u(x; c)$ for all nonnegative values of c.

We now consider briefly a possibility of applying this technique to the solution of the Dirichlet problem for the equation

$$\Delta u + cu = -g, \quad 0 < c < \lambda_1,$$

where λ_1 is the first eigenvalue of the operator $-\Delta/2$ for the considered domain D. It is known [9.5], that the solution to this problem satisfies an integral equation similar to (9.22), which differs by the kernel, namely, $d\sqrt{c}/\sinh(d\sqrt{c})$ is replaced with $d\sqrt{c}/\sin(d\sqrt{c})$. In addition, the weight function in the integral representing $h(r)$ is also changed using the corresponding Green function for the ball [9.5]. As shown in [9.5], the variance of the standard Monte Carlo estimate described above is finite if, for example, the diameter of the domain D is less than $\sqrt{2/c} \cdot \pi$. If this condition does not hold, then the standard Monte Carlo estimate is, in practice, not applicable. In this case, it is useful to construct vector Monte Carlo estimates for evaluating a finite part of (9.13) with corresponding operators K_1 and K_2, as mentioned in Section 9.3.

A. Models of Random Variables

We will present some standard and special methods for simulating random variables and vectors. Special attention is paid to the repetition method for constructing special random vectors with given one-dimensional distribution (the same for all components) and nonnegative correlation coefficients. In addition, this method is generalized to a wider class of random vectors [A.1–4].

A.1 Simulation of Random Variables

Simulation of a random variable ξ with a given distribution is carried out, as a rule, by transformations of one or a number of independent values of a random number α uniformly distributed between 0 and 1, i.e., by the following simulation formula:

$$\xi = \varphi(\alpha_1, ..., \alpha_n). \tag{A.1}$$

The sequences of samples of α are obtained usually by the multiplicative congruental method:

$$u_0 = 1,\ u_n = u_{n-1}5^{2p+1}(\mathrm{mod} 2^m),\ \alpha_n = u_n 2^{-m},$$

where 2^m is a large integer determined by the design of the computer, and $p = \max\{q : 5^{2q+1} < 2^m\}$. The numbers of this type are called a pseudorandom sequence. Such a sequence will repeat itself after at most 2^{m-2} steps. Note that in Sect. 8.2 we described a special pseudorandom simulation method which guaranties a certain correlation in Monte Carlo calculations. It is useful to apply the congruental method there.

For simulating a discrete random variable ξ with the distribution

$$\mathbf{P}(\xi = x_k) = p_k,\ k = 0, 1, ...,$$

one takes $\xi = x_m$, if

$$\sum_{k=0}^{m-1} p_k \le \alpha < \sum_{k=0}^{m} p_k.$$

The method of inverse functions is standard for simulating a continuous random variable. It is based on the following obvious relation:

$$\xi = F^{-1}(\alpha),$$

where F is the distribution function with the given density $f(x)$.

The composition method is based on the randomized simulation of the density

$$f(x) = \sum_k p_k f_k(x).$$

First, the number m is sampled from the distribution

$$\mathbf{P}(m = k) = p_k,$$

next ξ is simulated according to the density $f_m(x)$.

The rejection method (the Neuman's method) is based on the following statement.

If the point (ξ, n) is uniformly distributed in the domain

$$G = \{(x, y) : 0 \leq y \leq g(x)\},$$

then

$$f_\xi(x) = g(x)/|G|.$$

In the rejection method one samples a point (ξ_0, η) uniformly distributed in a domain $G_1 \subset G$, and one takes $\xi = \xi_0$ if $(\xi_0, \eta) \in G$; otherwise the choice of $\xi_0, \eta)$ is repeated, etc.

To sample the point (ξ_0, η) from the uniform distribution in the domain

$$G_1 = \{(x, y) : 0 \leq y \leq g_1(x)\} \quad ,$$

it is sufficient to sample ξ_0 from distribution $g_1(x)/G_1$ and take $\eta = \alpha g_1(\xi_0)$. The function $g_1(x) \geq g(x)$ must be chosen so that the random ξ_0 can be sampled by a standard method. For example, if $a \leq \xi \leq b$ and $g(x) = cf(x) \leq R$, then it is possible to take

$$\xi_0 = a + (b - a)\alpha_1, \ \eta = R\alpha_2.$$

The mean number of operations of the rejection method is proportional to $|G_1|/|G|$.

The quantity n is usually called the dimension of the simulation formula (1.1). The random variable $\xi = \varphi(\alpha_1, ..., \alpha_n)$ has the derived distribution, since the vector $(\alpha_1, ..., \alpha_n)$ is uniformly distributed in the n-dimensional cube. It is known that the uniformity property of the pseudonumbers is weakened as n increases. Therefore, it is useful to take n not so large in the simulation formulae.

The standard method of inverse function $\xi = F^{-1}(\alpha)$ is not often used in practice, since it is difficult to construct explicitly the function F^{-1}. We now present some special simulation formulae for a series of distributions.

1. Geometrical distribution with the parameter p:

$\xi = \text{entier}(\ln\alpha / \ln(1-p))$.

2. Discrete uniform distribution:

$\xi = \text{entier}(\alpha n) + 1$.

This formula can be applied for improving the simulation of arbitrary discrete distributions using a uniform stratification of probabilities.

3. Standard Gaussian distribution $N(0,1)$*:*

$$\xi = (-2\ln\alpha_1)^{1/2}\cos(2\pi\alpha_2),$$

$$\eta = (-2\ln\alpha_1)^{1/2}\sin(2\pi\alpha_2).$$

The random variables ξ, η are independent and have a Gaussian distribution with zero mean and variance 1.

4. Beta distribution with parameters p, q on the interval (0,1).

(a') $q = m$ is integer:

$$\xi = \exp\left\{\sum_{k=1}^{m} \frac{\ln\alpha_k}{p+k-1}\right\};$$

(a") $p = m$ is integer: after the change $\zeta = 1 - \xi$ we came to the case (a').

(b') p, q are not integers, and $q > 1$:

$$\xi = \exp\left\{\sum_{k=1}^{m+1} \frac{\ln\alpha_k}{\vartheta+p+k-1}\right\},$$

where $m = \text{entier}\ q$, and ϑ is an integer valued random variable with the distribution $\mathbf{P}(\vartheta = n) = p_n$, where the probabilities are recurrently calculated:

$$p_{n+1} = p_n \frac{(n+p)(n+1-m-q)}{(n+1)(n+1+m+p)}, \quad p_0 = \frac{B(p,m+1)}{B(p,q)};$$

(b") p, q are not integers, and $p > 1$: after the change $\zeta = 1 - \xi$ we come to the case (b').

(c) $p < 1, q < 1$: then the composition method is applicable, since

$$f_{pq}(x) = \frac{p}{p+q} f_{p+1,q}(x) + \frac{q}{p+q} f_{p,q+1}(x),$$

and we go to the cases (b') and (b").

5. Gamma distribution with the parameter ν*:*

$$\xi = -\ln\left(\prod_{k=1}^{m} d_k\right) - \zeta_\mu \ln\alpha_{m+1},$$

where $m = \text{entier}\ \nu$, and ζ_μ is the random variable with the Beta distribution with parameters $p = \nu - m$, $q = 1 + m - \nu$.

A.2 Simulation of Random Vectors

A standard algorithm for simulating a continuous random vector $\xi = (\xi_1, ..., \xi_n)$ consists of a sequential sampling of its components from the conditional distributions according to the representation

$$f_\xi(x_1, ..., x_n) = f_1(x_1) f_2(x_2|x_1) ... f_n(x_n|x_1, ..., x_{n-1}).$$

The rejection and composition methods can be obviously generalized to the multi-dimensional case. We now describe some special methods for simulating random vectors.

1. The Gaussian vector $\xi = (\xi_1, ..., \xi_n)$ can be simulated by the formula

$$\xi = A\eta + \mathbf{M}\xi,$$

where $\eta = (\eta_1, ..., \eta_n)$ is a set of standard independent random values, and A is a triangle matrix whose entries are calculated by formulae

$$a_{ij} = \frac{K_{ij} - \sum_{k=1}^{j-1} a_{ik} a_{jk}}{\sqrt{K_{jj} - \sum_{k=1}^{j-1} a_{jk}^2}}, \quad q \le j \le i \le n,$$

where $K_{ij} = M[\xi_i - m_i)(\xi_j - m_j)] with (m_i = \mathbf{M}\xi_i)$. Note that if the entries a_{ij} are calculated approximately (e.g., by the Monte Carlo method), it may appear that the calculated matrix K is not positive definite. In this case the matrix K can be improved as follows. First, a transformation matrix T is constructed such that

$$TKT^{-1} = \Lambda = \{\lambda_{ii}\}, \quad i = 1, ..., n,$$

where Λ is a diagonal matrix. Then a matrix $\Lambda^{(0)}$ is constructed by replacing in Λ the negative values λ_{ij} by $|\lambda_{ij}|$, and zero elements, by small positive value ε. The transformation gives

$$K^{(0)} = T^{-1} \Lambda^{(0)} T.$$

In what follows, we use for a_{ij} the entries of the matrix $K^{(0)}$.

2. A random vector with given one-dimensional distribution functions F_k and with a given correlation function can be constructed by the formulae

$$\xi_k = F_k^{-1}[\Phi(\eta_k)], \quad k = 1, ..., n,$$

where Φ is the distribution function of the standard Gaussian random value, and $(\eta_1, ..., \eta_n)$ is a vector of standard Gaussian random values with covariances satisfying the known system of equations. It may appear that this system is not solvable.

3. In [A.5], a method (the repetition method) for constructing random vector with positive correlations and a given one-dimensional distribution was presented. In each sample the components are arranged into groups of equal values such that the values of different groups are independent and distributed according to the one-dimensional distribution law.

From this we can formulate the following general principle for constructing the repetition method: for all variants Δ_m of the above subdivisions the probabilities p_m must be defined so that the random vector can have the desired correlation matrix. It is clear that for a fixed subdivision Δ_m a random vector with correlation matrix $R(\Delta_m) = \|r_{ij}^{(m)}\|$, whose entries are zero or unit elements such that $r_{ij} = 1$ if the components ξ_i, ξ_j belong to one group. Thus the quantities p_m are calculated from

$$R = \sum_{m=1}^{N} p_m R(\Delta_m), \tag{A.2}$$

where N is the total number of subdivisions.

For simplicity, we choose the indexation of the subdivision Δ_m according to the decrease of the number of units in matrices $R(\Delta_m)$. Then $R(\Delta_1)$ is a matrix which consists of units, while in $R(\Delta_N)$ the units are located only on the diagonal. Taking this into account, we rewrite (A.2) as a system of $n(n-1)/2$ algebraic equations

$$p_1 + \sum_{m=2}^{N-1} p_m r_{ik}^{(m)} = r_{ik},\ i < k, \tag{A.3}$$

where $p_m \geq 0,\ \sum\limits_{m=1}^{N} p_m = 1$.

If the solution with the mentioned properties exists, then the vector $(\xi_1, ..., \xi_n)$ can be simulated by the repetition method. Otherwise, such a simulation is not possible. The problem formulated can be solved as follows. Using standard methods of linear programming (e.g., the simplex method [A.6]), the nonnegative solution $(p_1^*, ..., p_{N-1}^*)$ of (A.3) having minimal sum of the components P^* is calculated. If $P^* \leq 1$, then we set $p_N = 1 - P^*$; thus, the random vector can be simulated by the repetition method according to the randomized representation (A.2). If $P^* > 1$, the repetition method is not applicable.

It should be noted that in the approach described the method used to sort out the variants of subdivisions and to calculate the quantities $r_{ij}^{(m)}$ [i.e., the matricies $R(\Delta_m)$] is very important. The total number of all subdivision variants is equal to the sum $\sum_{k=1}^{n} S(n,k)$, where $S(n,k)$ is the Stirling number

$$S(n,k) = \frac{1}{k!} \sum_{j=0}^{k} (-1)^j C_k^j (k-j)^n,$$

which describes the number of different variants of subdivisions of a set of n different elements into k parts. These numbers can be calculated by the recurrent formula

$$S(n+1,k) = S(n,k-1) + kS(n,k) \quad .$$

The following asymptotic formula holds: $S(n,k) \sim k^n/k!$.

In Table A.1 the sum $\sum_{k=1}^{n} S(n,k)$ for different values of n is shown:

Table A1

n	1	2	3	4	5	6	7	8	9	10
$\sum_k S(n,k)$	1	2	5	15	52	203	877	4140	21147	115675

Table A2

r	m			
	1	2	3	4
r_{12}	1	1	0	0
r_{13}	1	0	1	0
r_{23}	1	0	0	1

As Table A.1 shows, the direct calculation of p_m by standard methods of linear programming is not possible if n is large. We now formulate a simple algorithm for solving (A.3).

1. $n = 3$. The system (A.3) is defined by Table A.2.

Each column defines a matrix $R(\Delta_m)$. Simple analysis shows that the optimal solution (i.e., having a minimal sum of the components) is given by

$$p_1^* = \rho = \min(r_{12}, r_{13}, r_{23}),\ p_2^* = r_{12} - \rho,$$

$$p_3^* = r_{13} - \rho,\ p_4^* = r_{23} - \rho.$$

Consequently, the repetition method is applicable for simulating a three-dimensional vector iff

$$p^* = \rho + \sum_{i<k} (r_{ik} - \rho) < 1.$$

The corresponding algorithm can easily be constructed using Table A.2.

2. $n = 4$. The system (A.3) is defined by Table A.3.

The quantities p_m can be calculated by the following simple (but not optimal) method, which is analogous to the case $n = 3$.

Table A3

r	m													
	1	2	3	4	5	6	7	8	9	10	11	12	13	14
r	1	1	1	0	0	1	0	0	1	0	0	0	0	0
r	1	1	0	1	0	0	1	0	0	1	0	0	0	0
r	1	0	1	1	0	0	0	1	0	0	1	0	0	0
r	1	1	0	0	1	0	0	1	0	0	0	1	0	0
r	1	0	1	0	1	0	1	0	0	0	0	0	1	0
r	1	0	0	1	1	1	0	0	0	0	0	0	0	1

Step 1. Calculate

$$p_1 = \rho_1 = \min_{i<k} r_{ik}.$$

Rewrite (A.3) in the form

$$\sum_{m=2}^{N-1} p_m r_{ik}^{(m)} = r_{ik} - \rho_1 = \delta_{ik}^{(1)}.$$

If $p_1 = \rho_1 + \sum_{i<k}(r_{ik} - \rho_1) \leq 1$, then we set

$$p_9 = \delta_{12},\ p_{10} = \delta_{13},\ p_{11} = \delta_{14},\ p_{12} = \delta_{23},\ p_{13} = \delta_{24},$$

$$p_{14} = \delta_{34},\ p_2 = p_3 = p_4 = p_5 = p_6 = p_7 = p_8 = 0;$$

thus, the repetition method is defined.

Step 2. Calculate

$$p_2 = \rho_2 = \min(\delta_{12}; \delta_{13}; \delta_{23}),$$

and continue as in Step 1.

Furthermore, the solution is constructed by induction. On the s-th step we obtain p_m and the following system:

$$\sum_{m=s+1}^{N-1} p_m r_{ik}^{(m)} = \delta_{ik}^{(s)}.$$

If $P_s = \sum_{m=1}^{s} p_m + \sum_{i<k} \delta_{ik}^{(s)} < 1$, then the calculations are finished (as in Step 1); otherwise continue to the next step. It is easy to see that the quantities P_s do not increase. After the quantity p_{14} is calculated, the condition $\sum p_m \leq 1$ is verified. If this condition is not satisfied, the repetition method cannot be constructed in such a way.

As an example, we calculated p_m for $r_{ik} = r^{|i-k|}$, $(0 < r < 1)$:

$$p_1 = r^3, \quad p_2 = p_5 = r^2 - r^3, \quad p_6 = r - r^2,$$

$$p_{12} = r - 2r^2 + r^3, \quad p_3 = p_4 = p_7 = p_8 = p_9 = p_{10} = p_{11} = 0.$$

The sum of these quantities is equal to $1-(1-r)^2$. It is not difficult to see that the same value gives the minimal sum of quantities p_m for $n=3$, $r_{ik}=r^{|i-k|}$. It is clear that when adding components, the minimal sum p_m does not decrease. Thus the solution of the problem in this case is optimal. Additivity of the algorithm implies that the solution at $n=4$ will be optimal also for $r_{ik}=\sum_i q_l r_l^{|i-k|} (q_l \geq 0)$.

The algorithm described is directly generalized to arbitrary n. Note that in the solution many values of p_m are equal to zero. This simplifies the implementation of the repetition algorithm. The number of positive p_m is less than $n(n-1)/2$, since, after obtaining each nonzero value of p_m, one of the components of the right-hand side of (A.3) vanishes (the corresponding equation is eliminating).

The method of calculation of p_m described can be defined as a method of successive elimination of the unknown values corresponding to the decrease of the number of units in the matrices $R(\Delta_m)$. Qualitative arguments show that the obtained solution is close to an optimal one. However, there are no exact results in this field.

4. We now describe the generalized repetition method presented in [A.7]. Let us consider the following problem. Let r_{ij} be a positive definite $n \times n$-matrix, and let $F(x)$ be a distribution function. It is necessary to construct a random vector $(\xi_1, ..., \xi_n)$ such that

$$\begin{aligned} &P(\xi_i < x) = F(x), \ i = 1, ..., n; \\ &\rho(\xi_i, \xi_j) = r_{ij}, \ i, j = 1, ..., n. \end{aligned} \tag{A.4}$$

Here $\rho(\xi_i, \xi_j)$ is the correlation coefficient of random variables ξ_i and ξ_j. This problem [under the additional assumption that $F(x) = 1 - F(-x+0)$] will be called a main problem. In what follows $\eta_1, ..., \eta_n$ are independent random variables distributed with $F(x)$, and δ_i is either +1 or -1.

The repetition method has been suggested for solving this problem [A.5].The idea is to construct the solution $(\xi_1, ..., \xi_n)$ using the random values $\eta_1, ..., \eta_n$. Assuming that the random values ξ_i have symmetric distribution, we now analyze the possibility of constructing the solution using one random variable.

Let $p(i_1, ..., i_n; \delta_1, ..., \delta_n)$ be nonnegative numbers such that the indices take the values $1, 2, ..., n$, and the sum of $p(i_1, ..., i_n; \delta_1, ..., \delta_n)$ over all values of $i_1, ..., i_n$ and $\delta_1, ..., \delta_n$ is equal to 1. We define the random vector $(\xi_1, ..., \xi_n)$ as follows: we take $(\xi_1, ..., \xi_n) = (\delta_1 \eta_{i_1}, ..., \delta_n \eta_{i_n})$ with probability $p(i_1, ..., i_n; \delta_1, ..., \delta_n)$.

We conclude from the theorem of conditional probabilities that the distribution function of ξ_1 is $F(x)$, and

$$\rho(\xi_j, \xi_m) = \sum \delta_j \delta_m p(i_1, ..., i_n; \delta_1, ..., \delta_n) \rho(\eta_{i_j}, \eta_{i_m}), \tag{A.5}$$

where the sum is taken over all values of $i_1, ..., i_n$ and $\delta_1, ..., \delta_n$. The problem is thus reduced to finding a set of probabilities $p(i_1, ..., i_n; \delta_1, ..., \delta_n)$ such that the correlation coefficients (A.5) satisfy the conditions (A.4). This method of solving the main problem is called the generalized repetition method. A particular case for this method was described in [A.5], where it was supposed that $p(i_1, ..., i_n; \delta_1, ..., \delta_n) > 0$ iff $\delta_1 = ... = \delta_n = 1$. However, the generalized repetition method can not be directly applied to nonsymmetric distributions.

The generalized method has two advantages compared to the standard repetition method. First, it solves the problem when r_{ij} are negative. Second, if $r_{ij} \geq 0$, it solves a broader class of problems. For example, for

$$\begin{bmatrix} 1 & 1/2 & 1/2 & 0 \\ 1/2 & 1 & 1/2 & 1/2 \\ 1/2 & 1/2 & 1 & 1/2 \\ 0 & 1/2 & 1/2 & 1 \end{bmatrix},$$

there is no repetition method for solving the corresponding problem. However, this problem can be solved by the generalized repetition method with

$$\mathbf{P}(\xi_1 = \eta_1,\ \xi_2 = \eta_1,\ \xi_3 = \eta_1, \xi_4 = \eta_1) = 1/2,$$

$$\mathbf{P}(\xi_1 = \eta_1,\ \xi_2 = \eta_2,\ \xi_3 = \eta_3, \xi_4 = -\eta_1) = 1/2.$$

Let us consider another example. Let $n = 2$. In this case there are two different solutions:

$$\mathbf{P}(\xi_1 = \eta_1,\ \xi_2 = \eta_1 \operatorname{sign} r_{12}) = |r_{12}|,$$

$$\mathbf{P}(\xi_1 = \eta_1,\ \xi_2 = \eta_2) = 1 - |r_{12}|.$$

and

$$\mathbf{P}(\xi_1 = \eta_1,\ \xi_2 = \eta_1) = (1 + r_{12})/2,$$

$$\mathbf{P}(\xi_1 = \eta_1,\ \xi_2 = -\eta_1) = (1 - r_{12})/2.$$

In the last solution both random values ξ_1 and ξ_2 are constructed using a single random value.

This example poses a question. Is it always possible to construct the desired values $\xi_1, ..., \xi_n$ using a single random variable? The positive answer gives the theorem formulated below.

Let us give a definition. Let $p(\delta_2, ..., \delta_n)$ be nonnegative numbers such that

$$\sum_{\delta_2,...,\delta_n} p(\delta_2, ..., \delta_n) = 1.$$

We put $(\xi_1, ..., \xi_n) = (\xi_1, \delta_2\xi_1, ..., \delta_n\xi_1)$ with probability $p(\delta_2, ..., \delta_n)$, where the random variable ξ_1 is distributed according to $F(x)$. The main problem is to find $p(\delta_2, ..., \delta_n)$ such that the vector $(\xi_1, ..., \xi_n)$ satisfies condition (A.4). This technique is called a degenerate repetition method. In this case the correlation coefficients $p(\xi_i, \xi_j)$ are calculated by the formula

$$\rho(\xi_i, \xi_j) = \sum \delta_i\delta_j p(\delta_2, ..., \delta_n), \tag{A.6}$$

where $\delta_1 = 1$, and the sum is taken over all values of $\delta_2, ..., \delta_n$. It is clear that this method presents a particular case of the generalized repetition method.

Theorem. If the main problem can be solved by the generalized repetition method, then it can be solved by the degenerate repetition method as well.

Proof. Suppose that the main problem can be solved by the generation method. Note that if we replace the independent random variables $\eta_1, ..., \eta_n$ in this solution by the uncorrelated random variables $\nu_1, ..., \nu_n$ distributed with $F(x)$, then the new random variables $\xi_1, ..., \xi_2, ...$ also solve the main problem, as follows from (A.5).

As $(\nu_1, ..., \nu_n)$ we take a vector which is equiprobably sampled from 2^n random vectors $(\delta_1\eta_1, ..., \delta_2\eta_1, ..., \delta_n\eta_1)$. It is easy to verify that the random variables $\nu_1, ..., \nu_n$ are then uncorrelated, and the vector $(\xi_1, ..., \xi_n)$ depends only on η_1. Thus, the solution $(\xi_1, ..., \xi_n)$ of the main problem obtained by the generalized repetition method posesses the following property: $p(i_1, ..., i_n; \delta_1, ..., \delta_n) \gg 0$ only if $i_1 = ... = i_n = 1$.

Let

$$p(\delta_2, ..., \delta_n) = p(1, ..., 1; 1, \delta_2, ..., \delta_n) + p(1, ..., 1; -1, -\delta_2, ..., -\delta_n).$$

It follows from (A.5, 6) that the constructed probabilities $p(\delta_2, ..., \delta_n)$ are the desired solution of the main problem constructed by the degenerate repetition method. □

The proved theorem allows us to reduce the number of unknowns from $(2n)^n$ to 2^{n-1} in the generalized repetition method. Note that the further reduction of an unknown can be carried out by the method of linear programming [A.5].

Example. Assume that

$$1 + \sum_{i>j} \delta_i \delta_j r_{ij} \geq 0 \quad \forall(\delta_1, .., \delta_m), \tag{A.7}$$

then it is easy to verify that

$$p(\delta_2, ..., \delta_n) = 2^{-n+1} \left(1 + \sum_{i>j} \delta_i \delta_j r_{ij} \right). \tag{A.8}$$

$(\delta_1 = 1)$ is the solution to the main problem obtained by the degenerate repetition method. Note that at $n = 3$ the numbers (A.8) define the unique repetition method. Therefore, at $n = 3$, the condition (A.7) is necessary and sufficient to obtain the solution by the generalized repetition method.

We note in conclusion that the repetition method gives a random vector ξ with a special degenerate multi-dimensional distribution whose density (e.g., in $2-d$) has the form

$$rf(x)\delta(y-x) + (1-r)f(x)f(y).$$

If the distribution with density f(x) is infinite divisible, then the multidimensional distibution to be simulated can be improved by summing independent samples of $\xi^{(m)}$, as described in Sect. 6.3.

References

Chapter 1

1.1 S.M. Ermakov, G.A. Mikhailov: *Statistical Simulation* (Nauka, Moscow 1982) [in Russian]

1.2 R.E. Edwards: *Functional Analysis* (Holt, Rinehart and Winston, New York 1965)

1.3 K.K. Sabelfeld: *Monte Carlo Methods in Boundary Value Problems* (Springer, Berlin–Heidelberg 1991)

1.4 S.G. Mikhlin: *Lectures on Linear Integral Equations* (Fizmatgiz, Moscow 1959)[in Russian]

1.5 Yu.V. Bulavsky: Dokl. Akad. Nauk SSSR. 283, 797-800 (1985), [Engl.: Sov. Dokl. 32, 194-198 (1985)]

1.6 P. Billingsley: *Convergence of Probability Measures* (John Wiley & Sons, New York 1968)

1.7 A.S. Frolov, N.N. Chentsov: Sov. J. Num. Math. Math. Phys. 2 , 714-717 (1962) [in Russian]

1.8 J. Spanier, E.M. Gelbard: *Monte Carlo Principles and Neutron Transport Problems* (Addison-Wesley, Reading, PA 1969)

1.9 B. Davison: *Neutron Transport Theory* (Oxford University Press, 1957)

1.10 W.A. Coleman: Nucl. Sci. Engn. 32, 76-81 (1968)

1.11 G.I. Marchuk, G.A. Mikhailov, M.A. Nazaraliev, R.A. Darbinjan, B.A. Kargin, B.S. Elepov: *The Monte Carlo Methods in Atmospheric Optics* (Springer, Berlin– Heidelberg 1980)

1.12 G.W. Kattawar, G.N. Plass: Appl. Opt. 7,1519-1527(1968)

1.13 O.A. Makhotkin: *Solution of Radiative-Conductive Heat Transport Problems by Monte Carlo Method* (Computing Center, Novosibirsk 1980)[in Russian]

1.14 G.W. Brown: *Modern Mathematics for the Engineer*, ed. by E.F. Beckenbach (McGraw–Hill, New York 1956)

1.15 M.E. Müller: Ann. Math. Statistics. 27 , No 3 (1956)

1.16 A.V. Grechannikov: *Solution of Problems of Fluctuating Chemistry by Monte Carlo Method* Thes. Doct. Diss. Novosibirsk, 1983. [in Russian]

1.17 G.A. Bird: *Molecular Gas Dynamics* (Clarendon, Oxford 1976)

1.18 O.M. Belocerkovsky, V.E. Janitsky: Sov.J. Num. Math. Math. Phys. 1195-1208 (1975). [Engl. transl. Sov. J. Num. Math. Math. Phys. 15 (1975)]

1.19 F.G. Cheremisin: Sov. J. Num. Math. Math.Phys. 10, 654-665 (1970) [in Russian]

1.20 Yu.N. Kondjurin: Sov.J. Num. Math. Math. Phys. 26, 1527-1534 (1986) [Engl. transl. Sov. J. Num. Math. Math. Phys. 26 (1986)]

1.21 H. Kahn: "Use of Different Monte Carlo Sampling Techniques", in *Symposium on Monte Carlo Methods*, ed. by H.A. Meyer (Wiley, New York 1956) 146-190

1.22 J.M. Hammersley, D.C. Handscomb: *Monte Carlo Methods* (Methuen London 1964)

1.23 R. Bellmann: *Dynamic Programming* (Princeton University Press, Princeton, N. J. 1957)

1.24 G.E. Albert: "A General Theory of Stochastic Estimates of the Neumann Series for the Solution of Certain Fredholm Integral Equations and Related Series", in *Symposium on Monte Carlo Methods*, ed. by H.A. Meyer (Wiley, New York, 1956) 146-190

1.25 J.H. Curtiss: J. Math. Phys. 32, 209-232 (1954)

1.26 S.M. Ermakov, V.G. Zolotukhin: "Application of Monte Carlo Methods to Calculation of Radiation Protection" in *Questions of Physics of Radiation Protection* (Gosatomizdat, Moscow, 1963) 171-182 [in Russian]

1.27 G.A. Mikhailov: Sov. J. Num. Math. Math. Phys. 8, 590-599 (1968) [Engl. transl. Sov. J. Num. Math. Math. Phys. 8 (1968)]

1.28 G.A. Mikhailov, S.G. Musikhin: Sov. J. Num. Math. Math. Phys. 21, 432-440 (1981) [Engl. transl. Sov. J. Num. Math. Math. Phys. 21 (1981)]

Chapter 2

2.1 S.M. Ermakov, G.A. Mikhailov: *Statistical Simulation* (Nauka, Moscow 1982) [in Russian]

2.2 G.I. Marchuk, G.A. Mikhailov, M.A. Nazaraliev, R.A. Darbinjan, B.A. Kargin, B.S. Elepov: *The Monte Carlo Methods in Atmospheric Optics* (Springer, Berlin– Heidelberg 1980)

2.3 M.H. Kalos and P.A. Whitlock: *Monte Carlo Methods* Vol.1: Basics (John Wiley & Sons, New York 1986)

2.4 R. Bellmann: *Dynamic Programming* (Princeton University Press, Princeton, N. J. 1957)

2.5 J. Spanier, E.M. Gelbard : *Monte Carlo Principles and Neutron Transport Problems* (Addison-Wesley, Reading, PA 1969)

2.6 Yu.V. Bulavsky: Sov. J. Num. Math. Math. Phys. 24, 148-152 (1984) [Engl. transl. Sov. J. Num. Math. Math. Phys. 24 (1984)]

2.7 B. Davison: *Neutron Transport Theory* (Oxford University Press, 1957)

2.8 B.A. Kargin, A.V.Starkov: Sov. J. Numer. Anal. Math. Modelling 2, 1-14 (1987)

Chapter 3

3.1 G.I. Marchuk, G.A. Mikhailov, M.A. Nazaraliev, R.A. Darbinjan, B.A. Kargin, B.S. Elepov: *The Monte Carlo Methods in Atmospheric Optics* (Springer, Berlin– Heidelberg 1980)

3.2 G.A. Mikhailov: *Some Questions on Theory of Monte Carlo Methods* (Nauka, Novosibirsk 1974) [in Russian]

3.3 B.A. Kargin, A.V.Starkov: Sov. J. Num. Anal. Math. Modelling 2, 1-14 (1987)

3.4 G.H. Hardy, J.E. Littlewood and G.Polia: *Inequalities* (Cambridge 1934)

3.5 B. Davison: *Neutron Transport Theory* (Oxford University Press, 1957)

3.6 Yu.V. Bulavsky: Sov. J. Num.Math. Math. Phys. 24, 148-152 (1984) [Engl. transl. Sov. J. Num. Math. Math. Phys. 24 (1984)

3.7 M. Loeve: *Probability Theory* (D.Van Nostrand Company, Princeton, N. J. 1960)

Chapter 4

4.1 S.M. Ermakov, G.A. Mikhailov: *Statistical Simulation* (Nauka, Moscow 1982) [in Russian]

4.2 A. Wald: Ann. Math. Stat. 20 (1949)

4.3 R.E. Edwards: *Functional Analysis* (Holt, Rinehart and Winston, New York 1965)

4.4 P. Billingsley: *Convergence of Probability Measures* (John Wiley & Sons, New York 1968)

4.5 G.A. Mikhailov: Dokl. Akad. Nauk SSSR. 253, 1047-1050 (1980) (transl. Sov. Dokl. 22, 209-213 (1980))]

Chapter 5

5.1 G.W. Kattawar, G.N. Plass: Appl. Opt. 7, 1519-1527 (1968)

5.2 D.G. Collins, W.G. Blatter, M.B. Wells and H.G. Horat: Appl. Opt. 11, 2684-2696 (1972)

5.3 G.I. Marchuk, G.A. Mikhailov, M.A. Nazaraliev, R.A. Darbinjan, B.A. Kargin, B.S. Elepov: *The Monte Carlo Methods in Atmospheric Optics* (Springer, Berlin– Heidelberg 1980)

5.4 I.M. Sobol: "Monte Carlo Method for Calculating the Criticality in Multigroup Approximation " in *Monte Carlo Method in the Radiative Transfer Problem* (Atomizdat, Moscow, 1967) 232-254

5.5 G.A. Mikhailov: *Some Questions on Theory of Monte Carlo Methods* (Nauka, Novosibirsk 1974) [in Russian]

5.6 K.K. Sabelfeld: *Monte Carlo Methods in Boundary Value Problems* (Springer, Berlin–Heidelberg 1991)

5.7 S.M. Ermakov, G.A. Mikhailov: *Statistical Simulation* (Nauka, Moscow 1982) [in Russian]

5.8 G.A. Mikhailov: Dokl. Akad. Nauk SSSR. 253, 1047-1050 (1980) [Engl. transl. Sov. Dokl. 22, 209-213 (1980)]

5.9 M.SH. Birman et al.: *Functional Analysis* (Nauka, Moscow 1972)

5.10 H. Kahn : "Use of Different Monte Carlo Sampling Techniques", in *Symposium on Monte Carlo Methods*, ed. by H.A. Meyer (Wiley, New York 1956) 146-190

5.11 N. Danford, J.T. Schwartz:*Linear Operators. Part 1: General Theory* (Interscience Publishers, New York 1958)

5.12 G.A. Mikhailov: Sov. J. Num. Math. Math. Phys. 8, 590-599 (1968) [Engl. transl. Sov. J. Num. Math. Math. Phys. 8 (1968)

5.13 R. Bellmann: *Dynamic Programming* (Princeton University Press, Princeton, N. Y. 1957)

5.14 J.H. Holton: SIAM Rev. 12, 1-63 (1970)

5.15 N.N. Chentsov : Dokl. Akad. Nauk SSSR. 147, 45-48 (1962) [transl. Sov. Dokl. 147 (1962)]

Chapter 6

6.1 W. Feller: *An Introduction to Probability Theory and its Applications* Vol.2 (John Wiley & Sons,New York 1971)

6.2 G.A. Mikhailov: Izv. Acad. Nauk SSSR, Fizika Atm. i Okeana 18, 1289-1295 (1982) [Engl. transl. in Izv. Acad. Nauk SSSR, Fiz. Atm. i Okeana 18, 1982]

6.3 A.Ja. Khinchin:*Works in Mathematical Queueing Theory* (Fizmatgiz, Moscow 1963) [in Russian]

6.4 S.M. Ermakov, G.A. Mikhailov: *Statistical Simulation* (Nauka, Moscow 1982) [in Russian]

6.5 P. Billingsley: *Convergence of Probability Measures* (John Wiley & Sons, New York 1968)

6.6 V.S. Troinikov: Izv. Acad. Nauk SSSR, Fiz. Atm. i Okeana 20, 274-279 (1984) [Engl. transl. Izv. Acad. Nauk SSSR, Fiz. Atm.i Okeana 20, 1984]

6.7 J.L. Doob: *Stochastic Processes* (John Wiley & Sons, New York 1953)

6.8 B. Davison: *Neutron Transport Theory* (Oxford University Press, 1957)

6.9 K.K. Sabelfeld: *Monte Carlo Methods in Boundary Value Problems* (Springer, Berlin–Heidelberg 1991)

6.10 B.S. Elepov, A.A. Kronberg, G.A. Mikhailov, K.K. Sabelfeld: *Solution of Boundary Value Problems by Monte Carlo Methods* (Nauka, Novosibirsk 1980) [in Russian]

Chapter 7

7.1 H. Kahn : "Use of Different Monte Carlo Sampling Techniques", in *Symposium on Monte Carlo Methods*, ed. by H.A.Meyer (Wiley, New York 1956) 146-190

7.2 V.N. Ogibin: "On Application of "Splitting Technique " in Calculations of Particle Transfer by Monte Carlo Method" in *Monte Carlo Methods in the Radiative Transfer Problems* (Atomizdat, Moscow 1967) 72-82 [in Russian]

7.3 Yu.V. Bulavsky: "On the Choice of the Survival Probability in the Neumann–Ulam Scheme", in *Methods and Algorithms of Statistical Modelling* (Computing Center, Novosibirsk 1983) 13-20

7.4 M.H. Kalos: Nucl. Sci. Eng., 16, 227-234 (1963)

Chapter 8

8.1 M.H.Kalos: Nucl. Sci. Eng., 16, 227-234 (1963)

8.2 S.M. Ermakov, G.A. Mikhailov: *Statistical Simulation* (Nauka, Moscow 1982) [in Russian]

8.3 G.I. Marchuk, G.A. Mikhailov, M.A. Nazaraliev, R.A. Darbinjan, B.A. Kargin, B.S. Elepov: *The Monte Carlo Methods in Atmospheric Optics* (Springer, Berlin– Heidelberg 1980)

8.4 B.A. Kargin, S.V. Kuznetsov, G.A. Mikhailov: Izv. Acad. Nauk SSSR, Fiz. Atm. i Okeana 15, 1027-1035 (1979) [Engl.transl. Izv. Acad. Nauk SSSR, Fiz. Atm. i Okeana 15 (1979)]

8.5 B. Davison: *Neutron Transport Theory* (Oxford University Press, 1957)

8.6 P. Billingsley :*Convergence of Probability Measures* (John Wiley & Sons, New York 1968)

Chapter 9

9.1 G.I. Marchuk, G.A. Mikhailov, M.A. Nazaraliev, R.A. Darbinjan, B.A. Kargin, B.S. Elepov: *The Monte Carlo Methods in Atmospheric Optics* (Springer, Berlin– Heidelberg 1980)

9.2 A.S. Frolov and N.N.Chentsov : Sov. J. of Num. Math. Math. Phys. 2 , 714-717 (1962) [in Russian]

9.3 N. Danford, J.T. Schwartz: *Linear Operators. Part 1: General Theory* (Interscience Publishers, New York 1958)

9.4 G.A. Mikhailov: Sov. J. Num. Math. Math. Phys. 6, 380-384 (1966) [Engl. transl. Sov. J. Num. Math. Math. Phys. 6 (1966)

9.5 B.S. Elepov, A.A. Kronberg, G.A. Mikhailov, K.K. Sabelfeld: *Solution of Boundary Value Problems by Monte Carlo Methods* (Nauka, Novosibirsk 1980) [in Russian]

Appendix

A.1 D.H. Lehmer: *Mathematical Methods in Large-Scale Digital Calculating Machinery* (Harvard University Press, 141-146, 1951)

A.2 B. Jansson: *Random Number Generators* (Almquist & Wiksell, Stockholm 1966)

A.3 G.A. Mikhailov: *Some Questions on Theory of Monte Carlo Methods* (Nauka, Novosibirsk 1974) [in Russian]

A.4 G.E. Box, M.E. Muller: Ann. Math.Stat. 29, 610-611 (1958)

A.5 G.A. Mikhailov: Theory of Probability and its Applications, 19, 873-878 (1974) [in Russian]

A.6 G.B. Dantzig: *Linear Programming and Extension* (Princeton University Press, Princeton, N. J. 1957)

A.7 A.I. Sakhanenko: "On Degenerate Repetition Method" in *Monte Carlo Methods in Computational Mathematics and Mathematical Physics* (Computing Center, Novosibirsk 1974) 42-46

Subject Index

S. M. Rytov, Y. A. Kravtsov, V. I. Tatarskii

Principles of Statistical Radiophysics 1

Elements of Random Process Theory

1987. X, 253 pp. 28 figs.
Hardcover ISBN 3-540-12562-0

Principles of Statistical Radiophysics is a four-volume series that introduces the newcomer to the theory of random functions. It aims at providing the background necessary to understand papers and monographs on the subject and to carry out independent research in fields where fluctuations are of importance, e.g. radiophysics, optics, astronomy, and acoustics.
"Elements of Random Process Theory", the first volume, contains the essential mathematical prerequisites and definitions related to this topic. It deals in particular with the physics of random pulse processes, shot and flicker noises, fluctuations in self-oscillatory systems, random actions on linear and nonlinear discrete dynamical systems, Markov processes and stochastic differential equations.

S. M. Rytov, Y. A. Kravtsov, V. I. Tatarskii

Principles of Statistical Radiophysics 2

Correlation Theory of Random Processes

Translated from the Russian by A. P. Repyev

1988. X, 234 pp. 54 figs.
Hardcover ISBN 3-540-16186-4

Contents: Fundamentals of Correlation Theory. - Applications of Correlation Theory. - Spectral Theory of Random Actions on Dynamic Systems. - Certain Kinds of Nonstationary Processes. - References. - Subject Index.

S. M. Rytov, Y. A. Kravtsov, V. I. Tatarskii

Principles of Statistical Radiophysics 3

Elements of Random Fields

1989. X, 239 pp. 36 figs.
Hardcover ISBN 3-540-17829-5

Contents: Fundamentals. - Radiation and Diffraction of Random Wave Fields. - Thermal Electromagnetic Fields. - Single Scattering Theory. - References. - Subject Index.

S. M. Rytov, Y. A. Kravtsov, V. I. Tatarskii

Principles of Statistical Radiophysics 4

Wave Propagation Through Random Media

Translated from the Russian by A. P. Repyev

1989. X, 188 pp. 43 figs.
Hardcover ISBN 3-540-17828-7

This Volume is concerned with Markov random fields, Feyman diagrams in perturbation theory, wave propagation in media with large random inhomogeneities, multiple scattering and scattering by rough surfaces.